iTake-Over

iTake-Over

The Recording Industry in the Digital Era

David Arditi

ROWMAN & LITTLEFIELD
Lanham • Boulder • New York • London

Published by Rowman & Littlefield
A wholly owned subsidiary of The Rowman & Littlefield Publishing Group, Inc.
4501 Forbes Boulevard, Suite 200, Lanham, Maryland 20706
www.rowman.com

16 Carlisle Street, London W1D 3BT, United Kingdom

Copyright © 2015 by David Arditi

British Library Cataloguing in Publication Information Available

Library of Congress Cataloging-in-Publication Data

Arditi, David, 1982–
iTake-over : the recording industry in the digital era / David Arditi.
pages cm
Includes bibliographical references and index.
ISBN 978-1-4422-4013-1 (cloth : alk. paper) – ISBN 978-1-4422-4014-8 (ebook)
1. Sound recording industry. 2. Music and the Internet–Economic aspects. I. Title.
ML3790.A76 2015
384–dc23
2014033013

♾™ The paper used in this publication meets the minimum requirements of American National Standard for Information Sciences Permanence of Paper for Printed Library Materials, ANSI/NISO Z39.48-1992.

Printed in the United States of America

Contents

List of Figures

Preface

Despite talk of doom and gloom for the music business, the case for optimism—and the prospects for massive growth in the overall music ecosystem—is strong. (Csathy 2014)

For the recording industry, digitization means change. Old music formats are obsolete; the MP3 has replaced the compact disc (CD). There are no longer price barriers to listening to music; free streams or $0.99 digital singles have replaced $15 CDs. Music distribution happens over networks instead of at record stores; music listeners can instantaneously access nearly all music via iTunes rather than spending time driving to the mall, finding a parking space, and digging through album crates, only to find that the store does not carry a particular album. These changes have rocked the recording industry and the broader music industry over the better part of two decades.

However, change has been the constant for the recording industry as each new technology creates an opportunity to sell old music on new media formats. Whether we discuss the change from 78-rpm records to 33 1/3-rpm records; mono to stereo; AM radio to FM radio; records to 8-track cartridges; 8-tracks to tape cassettes; tapes to CDs (and everything in between)—the recording industry adapted to new media, often propelling the change forward. In fact, selling albums in new formats to people who already own an album—called the album replacement cycle—is an important revenue stream for major record labels. So the recording industry always plays a role in format transformations.

In 2014, the recording industry is looking to the next transformation to encourage the album replacement cycle by developing the MP3's successor. Audiophiles have never been fans of MP3s because the compression used to create these small files degrades sound quality (Sterne 2012). Since the MP3 was always a product of the constraints of limited Internet bandwidth and hard-drive space, the format becomes obsolete with faster Internet and cheaper, larger computer storage. As a result, the recording industry is beginning to promote what is called high-resolution audio (HRA). *Billboard* magazine, the trade journal of the recording industry, touts HRA as "a new format [that] gives rights owners a chance to resell yet another copy of their catalog titles" (Pham 2014). In addition to getting people to buy music they already own, the article further speculates that labels "have the opportunity to add a premium" for music in HRA (Pham 2014), that is, sell the same album for more money. The logic of the album replacement cycle is evident here as the recording industry advocates the repurchasing of albums that people already own in addition to a system of premium pricing.

What is interesting here is that the industry is going through a transformation similar to the change around the turn of the twenty-first century. In 1999 and 2000, the market for replacing vinyl records and tape cassettes with CDs slowed because catalog[1] sales plummeted (Park 2007). Since people can rip CDs onto hard drives, there was no incentive to repurchase albums in digital format. Following heavy marketing campaigns by iTunes and the major record labels in 2009, some deep catalog albums saw a surge in sales reminiscent of previous moments in the album replacement cycle. For instance, the release of the Beatles' entire catalog on iTunes, with heavy television advertising, created a rise in the sale of the Beatles' music (Bruno 2010b). Fast-forward to 2013 and the industry experienced a similar phenomenon as the "bottom has fallen out of catalog sales" for both digital albums and tracks (Christman 2014a, 34). The digital music track is going the way of the 8-track—it is becoming obsolete as people embrace new forms for listening to music.

Despite the noticeable decline of digital albums and tracks for both catalog and current music, industry-driven rhetoric demonstrates the renewed strength of the recording industry. As the title of a *Rolling Stone* article reflecting on 2013 proclaims: "2013: The Future Has Arrived"; the article declares that "2013 was the year the music industry tried to go from hot mess to strategic hot mess" (Hiatt 2013, 15). In other words, the recording industry went from over a decade of disarray from piracy to an industry in chaos because of celebrities vying for attention.[2] The idea behind the strategic hot mess was that blockbuster celebrities can sell more music at higher profits than breaking new artists (Elberse 2013). But even more important than the already-known idea of getting more revenue out of blockbuster celebrities was the increase in streaming and subscription services.

Streaming music became *the* alternative to physical music in 2013. "Spotify is now the No. 2 digital retailer for most labels in terms of income. . . . All told, streaming revenues make up approximately one-quarter of all income for most content owners, having quadrupled in the last two years" (McDaniels 2014). In fact, streaming subscription services have the major labels salivating over increased revenue as the "average consumer spends about $40 per year on recorded music. If the average consumer signs up for a streaming service, he or she will spend as much as $120 per year ($10 per month times 12 months)" (McDaniels 2014). A dramatic shift toward subscription services will induce a permanent rise in revenue that is completely separated from actual music consumption.

What we see in 2014 is a streamlined industry. If we use 1999 as the beginning of public transition to digital music, the industry went from six major record labels to three during the digital transition. This has resulted in numerous layoffs as large corporate labels purge their artist rosters and eliminate label employees to improve efficiency and profits, a process that has been going on since at least the 1960s (Chapple and Garofalo 1977). When two companies merge, there is duplicated labor—one way to increase profit is to eliminate this redundant labor.[3] Yet this says nothing about the effects of disintermediation—the removing of intermediaries in the process of distributing media (a process discussed throughout this book). Today, the three major labels consist of Sony Music Entertainment, Warner Music Group, and Universal Music Group. These three labels produce, distribute, and sell the vast majority of music.

Since recorded music is so tightly controlled by an oligopoly—an industry dominated by a small number of companies—the labels are constantly looking for new ways to create larger releases. Jay Z's *Magna Carta Holy Grail* was one of the most innovative releases of 2013. In a deal with Samsung, Jay Z agreed to release the album free to one million people exclusively through an app on Samsung's Galaxy smartphones on July 4, 2013, five days before its official release on July 9 (Knopper 2013). Samsung paid Jay Z and his label group (Roc-A-Fella, Roc Nation, and Universal) five dollars wholesale on one million albums in advance of the release for the right to this exclusive deal (Pham, Hampp, and Christman 2013). Since Samsung paid for the one million albums in advance, Jay Z recorded one million album sales regardless of the number of albums actually downloaded through Samsung's app, making it the first album to go platinum in one day.[4] The estimated $30 million investment by Samsung is the kind of "blockbuster bet" (Elberse 2013, 19) that large corporations make on entertainment celebrities. However, this high-risk investment is only conceivable in a digital music environment. Distributing music on smartphones was not possible in the 1990s. An exclusive Samsung deal in the 1990s would have involved distributing CDs to Samsung retail stores (which did not exist) throughout the United States. The one million unit shipments made to these retail outlets would have carried a twelve-dollar wholesale price (Hull, Hutchison, and Strasser 2011, 255), and at more than double the price, this investment would have looked much different to Samsung. However, without the cost of printing CDs and distributing them to stores, the deal is not only cheaper but also more profitable.

Following in the footsteps of Jay Z, Beyoncé shocked the world on December 13, 2013, with the surprise release of her self-titled album exclusively on iTunes. With no promotion, the secret release sold over 820,000 albums in three days, a new record for iTunes (Subramanian 2013). While this release strategy was largely successful, it came with two major consequences. First, Target and Amazon refused to carry the physical CD days before Christmas because they felt abandoned by the exclusive deal with iTunes (Christman 2014b); this likely reduced impulse buys at Target during the busiest shopping season. Second, the surprise album eliminated the opportunity for Beyoncé to release singles on the radio ahead of the album's launch, and a single from the album has yet to reach number one on Billboard's Hot 100. Once again, if we went back to the 1990s, this release would have been impossible because there would have been no way to keep the release secret. Even if Beyoncé released the album exclusively with one retailer in the 1990s—for example, Tower Records—hundreds to thousands of employees from production to retail would have been aware of the album. Furthermore, there would have been no way to predict the number of albums to ship to Tower Records, word of mouth could not have spread so quickly, and in some areas people would not have been able to buy the album because they lived too far from a Tower Records. iTunes allowed Beyoncé's management team to keep the album undisclosed because few people at iTunes needed to know about the forthcoming album.

The recording industry is very enthusiastic about the synergistic opportunities available from licensing music in video games. *Grand Theft Auto V*'s release on September 17, 2013, set the record for the largest first-day release of an entertainment product, hauling in $800 million in sales—and exceeding one billion dollars in one week. Record executives are excited because *Grand Theft Auto* "licensed 240 tracks and commissioned original songs from A$AP Rocky, Flying Lotus, Twin Shadow, Neon Indian, Yeasayer, OFF! and Tyler, the Creator, among others" (Pham 2013). These songs not only earn the recording artists and their labels revenue through synchronization licenses but also drive album and singles sales, digital streams, and concert attendance. Video games provide opportunities for record labels that did not exist before digital music.

The new media ecology has major record executives looking bright eyed to the future. With new prospects for consumers to replace their MP3s with files having better sound quality, an unlimited stream of consumption through subscription services, and a new vigor to use synchronization licenses, the streams of revenue and profit continue to increase for the major record labels.

Of course, this is not the story that the recording industry, led by the Recording Industry Association of America (RIAA), told from 1999–2010. On the contrary, the RIAA told a bleak story about the prospects for music: rampant piracy would lead to the death of music. How did we go from the death of music to a very vibrant recording industry? *iTake-Over* explores the political, economic, cultural, social, and technological shift from physical to digital music in the moment between 1995 and 2010. The music industry has experienced significant changes in the production, distribution, and consumption of music with the introduction of the Internet and digital technologies over the last two decades. This book demonstrates that the major record labels actively changed how they produce music with new technologies to remain dominant within the broader music industry. The availability of digital distribution transformed music production by compelling the recording industry to create digital music that they could profit from in new ways. Digital production and distribution systems enhanced, rather than undermined, the commercial position of these firms, which counters claims made by the recording industry, reflected in "piracy panic narratives," that they are experiencing a financial crisis.

Acknowledgments

My debts for this project run deep—too deep to name everyone who contributed to this book—but thank you to you all. This project began as my dissertation in George Mason University's cultural studies program under the direction of Paul Smith. Paul, Timothy Gibson, and Char Miller deserve my deepest gratitude as members of the dissertation committee. Their input and suggestions helped shape this project from an early stage—thank you. Jennifer Miller, Richard Otten, Ariella Horwitz, and Scott Killen, thank you for providing valuable insights and criticisms on drafts. Since this was my first time swimming the waters of book publishing, I would like to thank Roger Lancaster, Donna Akers, Robert Bing, and Maria Martinez-Cosio for their advice. I would also like to thank Bennett Graff for his interest in this project and help through the process.

Introduction

When the Recording Industry Association of America (RIAA)[1] filed a lawsuit against Napster, Inc.—a peer-to-peer (P2P) file-sharing service—in 1999 for alleged copyright infringement and unfair competition, the recording industry was staking a position that file sharing online, and digital music in general, was the death knell of the recording industry unless something was done to stop it.[2] Napster permitted users to download and upload files from each other free of charge and with no expectation for reciprocity, but many of these files were of copyrighted music. Within the industry's narrative about its impending death was a claim that the major record labels and their recording artists were the victims of widespread piracy and theft. However, not only has this collapse failed to come to fruition, but according to my research, the move to digital music appears to have strengthened the major record labels within the broader music industry. The major record labels seem to be stronger because they are selling more units,[3] are relying more on performance rights,[4] and, upon closer examination of the political economy of the music industry, appear to have increased their profit margins. Most importantly, in the process of trying to close down all digital competition, the RIAA and the major record labels have fundamentally changed the laws and norms that regulate cultural production.

To create the political will for the state to regulate cultural production, the RIAA created what I call a "piracy panic narrative" in which the recording industry claims to be experiencing financial turmoil. Industry representatives contend that there has been a decline in sales. Then they link this decline to consumers who have obtained free music online through file-sharing programs such as Napster, Kazaa, and Grokster. Finally, the industry's undemonstrated declaration of sales decline is then relegitimated in the press. For instance, in a 2003 *Billboard*[5] article, former RIAA president Cary Sherman asserts that the RIAA and the recording industry "believe the use of these illegal peer-to-peer services is hurting the music industry's efforts to distribute music online in the way consumers demand" (Garrity and Christman 2003). This belief is unsubstantiated in the article; throughout this book, I demonstrate that unsubstantiated claims form the bedrock of the music industry's piracy panic narratives. Moreover, data that I introduce demonstrate that the RIAA remains a very influential trade organization and that the recording industry continuously finds creative ways to profit from new technologies instead of being financially destroyed by them.

Claims like Sherman's construct a belief among the public that something must be done to stop file sharing or else the music industry as we know it will cease to exist. For example, *Billboard* accepts Sherman's declaration as evidence of decline and legitimates the industry's assertion that major record labels might have to cut costs to survive the crisis. However, there is no analysis of sales data. The article goes on to say that "labels will further re-evaluate head count, as well as other elements of their cost structures and business practices, as sales continue to struggle" (Garrity and Christman 2003). In other words, there will be layoffs because people share files. The causal link between declining sales and file sharing is validated by a type of superficial journalism that replaces investigation and hard evidence with press releases, which become evidence. In effect, press releases become fact because there is no counterview expressed within the articles about the decline of the recording industry.[6]

This book interrogates the causal link between file sharing and the decline of recorded music sales asserted by the RIAA. Additionally, I argue that the logic of capitalism is the only explanation for the aforementioned jobs cuts. As new technologies are developed, businesses eliminate labor at all levels of the production process in order to increase profits. Throughout this book, I emphasize the similarities between the recording industry's change of business models, its previous transformations, and shifts in other industries. This is not to discount the fact that CD sales have plummeted, but rather to express that the recording industry induced the unraveling of CDs as the main medium for music consumption because labels appropriate technological changes for their benefit. Without question, consumers buy far fewer CDs today than they purchased in 2000.[7] However, people in 2010 purchased far more digital music files than they purchased CDs at their zenith.

What is the significance of CD sales plummeting? The shift from CDs to digital music is part of a history of shifts that goes back to the transition from sheet music to piano rolls. With the rise of new technology that increases the sound fidelity and portability of recordings, the recording industry has always adjusted to sell its copyrighted music on each new medium. Consumer electronics manufacturers developing a new technology usually start these shifts — sometimes in opposition to the recording industry's wishes. However, the recording industry ultimately benefits from the change in media by reselling music to consumers in new media formats; this allows consumers to listen to the same music on different players. Eventually, major record labels accelerate these shifts for their own benefit. For instance, in the late 1980s, record labels forced retailers to stop carrying tape cassettes by refusing to buy back unsold tapes. The transition away from CDs is just the latest transmediation for the recording industry.

Along with the media shift, there is a generational cultural shift in the consumption of music. Simply put, kids download music—whether they use iTunes or a P2P program or stream music on YouTube. Now MP3s and other digital formats are the primary delivery media of music. Younger music listeners adapted to the idea that they *use* music instead of *owning* it. Meanwhile, older music listeners lament the demise of the CD or romanticize the "warmth" of vinyl records. Furthermore, consumers worry about the impact of digital singles on full-length albums in the short term and the vitality of music in the long term.

Over the past two decades, the music industry has seen profound shifts in its political economy because of the digital media distribution made possible by information and communications technologies (ICTs). Between 1995 and 2010,[8] a significant transformation occurred in the music industry as the primary recorded commodity shifted from CDs to digital music files (IFPI 2009). Throughout this book, I call this shift the "digital transformation of the recorded music commodity" or, alternatively, the "digital transformation." During this transformation, major record labels have not only relied more on digital files to distribute music, but they have exploited digitization to increase the utilization of performance copyrights through cross-promotion on television, in video games, and in movies. This digital transformation is significant because as the recorded commodity adapts to digital media, disintermediation—the elimination of intermediaries in production and distribution chains—alters the relations of production in the music industry. While the Internet presents musicians and fans with new ways to interact, major record labels also exercise their dominant position in the music industry to their benefit. In other words, far from being passive victims of technological shifts in the recorded commodity form, the RIAA has been an active player in creating novel ways to profit from new modes of commodification, and it has used the change in commodity form to consolidate major record label power to get the public and the state to invest in "saving" music.

I stress that transformations of technology do not happen in a vacuum.[9] All technological change happens as part of a social process, so it is imperative to analyze that process in order to understand the impact that technology has on society. By studying the social processes that lead to technological change, we can see the influence previous technologies and ideologies[10] about technology had on the formation of emergent technology. It is essential to recognize that emergent technologies "are not pregiven but the result of the dynamical [*sic*] development of complex systems" (Fuchs 2008). Intellectual property laws are a part of the complex system that helps to determine the form and usage of emergent technologies. Where copyright owners historically deployed intellectual property to regulate how individuals reproduce copyrighted material, new policies restrict the actual use of content. These policies have been especially significant for structuring the way people consume media through the Internet.

The digital transformation of the recorded commodity happened as part of a complex social process affected by economic, political, and cultural forces comprising a total social field, which makes the adaptation of digital technology dialectical. The digital transformation of the recorded commodity is dialectical because it is both a reaction to and a preemptive evolution of technology by the recording industry. It is an oversimplification to argue that Napster and MP3s surprised the recording industry; at the same time, without these innovations, there would be only marginal reasons for major record labels to look beyond the CD. Rather, the emergence of MP3 files gave the major record labels an incentive to change their production and distribution processes because their standard business models were quickly becoming obsolete. New business models for the recording industry promised to change the social relations of production by eliminating the need for specific types of labor and material goods. In short, digital technology and the Internet provided major record labels with the impetus to change their business models and cut labor costs.

If we look at the broader relations of production that changed because of neoliberalism and digital networks, it becomes clear that digital media and distribution were a logical outcome of creating a more efficient capitalism. Rather than viewing the Internet as a revolutionary new communication system started by geeks in garages, a more sensible approach is to analyze the way that government policies have changed to advance the needs of corporate oligopoly. For instance, in *Digital Capitalism*, Dan Schiller argues that the Internet was a network consciously developed by reorganizing "telecommunications policy on neoliberal lines" (2000, 203).[11] Schiller contends that this global privatized network was necessary, along with free-trade policies, to deliver global capitalism efficiently. In this respect, digital retail allows record labels to deliver digital music anywhere through the Internet, which eliminates costly physical distribution. Along with cutting the cost of manufacturing (no longer printing CDs) and distribution (gasoline, trucks, shipping containers, etc.), this system of disintermediation also eliminates labor at numerous points in the commodity chain. Digital distribution, in turn, has allowed fewer workers to produce more commodities that can reach remote points around the world.

In the recording industry, the ability to preserve state-sanctioned intellectual property monopolies in the new digital information environment, while effective, is fragile because the same technologies that allow major record labels to cut costs also allow consumers to obtain free music. To counter the ability of consumers to circumvent the major labels' digital profit system, the recording industry constructed a narrative that positions all file sharers as criminals. In other words, they launched a war on what they viewed as poaching.

These shifts have involved changes in the production, consumption, and distribution of music; the conception of ownership of intellectual property; and the way that labor works in the music industry. While these shifts have occurred in most industries, the music industry is important because (1) there has been a significant amount of discourse in the news about how digital distribution affects the music industry, and (2) the music industry was one of the first industries to be directly affected by Internet distribution and digital technologies. As the music industry has responded to these transformations, the fundamental commodity of the music industry has shifted along with the relationship between capital and labor. The overarching question behind this book is what effects has the digital transformation of the music commodity had on the relations of production[12] in the music industry?

Rather than finding that the recording industry was *reacting to* the digital transformation, I reveal that the actions of major record labels and the RIAA *led to* the digital transformation. While there were times when alternative means of distribution developed outside of the recording industry (Napster is the exemplary case of this), even the opportunities that those means of distribution presented had been planned for by the recording industry; this is clear from the RIAA's legislative strategies during the 1990s. Throughout this book, I illustrate that over the better part of the past two decades, major record labels and the RIAA proactively altered the relations of production and the record labels appear to have strengthened (financially and politically) their dominant position within the broader music industry. My argument thus counters the recording industry's narrative—a narrative that constructs the industry as victims of "pirates" who refuse to play by the rules. The digital transformation has been a game changer, but it is a game the industry continues to win by creating the rules.

The main findings of this book appear counterintuitive because of the commonsense understanding of the political economy of the recording industry. However, upon closer examination of the recording industry's narrative of financial hardship, there is reason to be skeptical of this narrative.

PIRACY PANIC NARRATIVE

"Would you go into a CD store and steal a CD? It's the same thing, people going into the computers and logging on and stealing our music."—Britney Spears (Quoted in Ahrens 2002)

The suggestion that file sharing is equivalent to stealing music has been the focal point of the recording industry's contention that the practice must be stopped for recorded music to continue to exist. I call this argument the "piracy panic narrative." This narrative's structure follows a pattern: file sharing is piracy, piracy is stealing, and stealing hurts artists and their labels. Therefore, major record labels argue that music fans who share files are not listening to free music but rather stealing income from their favorite artists.

According to Britney Spears, downloading music through file-sharing websites is stealing. Spears's statement was part of an advertising blitz at the apex of the Recording Industry Association of America's battle against file sharing in 2002. This ad blitz included one-page ads in newspapers across the country and TV and radio commercials. "Nearly 90 singers and songwriters," one *Washington Post* article claims, "have signed the newspaper ad, and several have lent quotes to the campaign" (Ahrens 2002). Additionally, the ad blitz itself generated its own news after reporters wrote stories about it, many without any counterbalancing perspective on the rhetoric. This argument is not only advanced by the recording industry but also by the news media, which become the platform for the recording industry's narrative. For their part, journalists give credibility to the piracy panic narrative by selling this narrative to the public and presenting the RIAA's press releases as news while rarely providing counterarguments to the industry's position. For instance, in one *Washington Post* news article, Rob Pegoraro asserts that the recording industry "pointed with justified alarm to a culture of theft on file-downloading services" (2002). Most of these stories unquestionably supported the RIAA's position. Throughout this book, I scrutinize the different ways the recording industry deploys this argument to justify changes to the relations of production in the broader music industry.

This ad campaign was part of a broader piracy panic narrative that began with the popularization of Napster in 1999 and continues today. The piracy panic narrative is a "calculated political strateg[y] to psychologically demonize opponents to make them appear to be 'bad' people. Because these bad people are doing bad things, they must be punished the way bad people are: by being sued, by paying exorbitant damages, and in some cases by going to jail" (Patry 2009, 44). By labeling file sharers as property thieves, the piracy panic narrative constructs a victim—the artist—and a victimizer—the fan; this pits musicians directly against their audiences, fans and consumers—that is, the people who always already financially support these musicians.

The problem is that statements such as Spears's are purely rhetorical. William Patry argues in *Moral Panics and the Copyright Wars* (2009) that by constantly repeating piracy metaphors, copyright-based industries attempt to do more than reframe the debate; they try to associate file sharing with stealing permanently. Downloading music from peer-to-peer file-sharing programs is not the same thing as stealing; in fact, legally, it is not even property theft. Copyright law is a "regulatory privilege" (Patry 2009, 110), not a form of property law; it cannot be compared to property theft because when a user downloads music, they are not taking something away from another user. The original user still has the ability to listen to the downloaded music and can still allow others to download their music. File sharers are not stealing music.

Of course, this is only one dimension of the recording industry's argument; the RIAA tries to go deeper into the theft analogy by monetizing music. Part of this argument is that if consumers pay for a CD, they are paying a recording artist to listen to their work, but if that same person does not pay for music, they are refusing to pay that recording artist for their work. This argument again contains faulty logic and is problematic for two reasons. First, it conflates the act of listening to music with a need to pay for music. Second, it ignores the role of record labels in profiting from the labor of many recording artists without compensating them.

Digital file sharing is an apex in the discourse about copyright policy, but this fear of piracy goes back further. Since tape cassette players became available with a "record" button,[13] major record labels have claimed that new technologies enable their users to make copies easily, allowing them to violate copyright law. Copyright owners (in this case, major record labels) contend that these technologies in effect enable piracy. With the development of each new technology, copyright owners assert that consumers will reproduce and distribute copyrighted content. Major record labels are careful not to recognize nuance in the actual reproduction and distribution practices of their consumers; rather, the recording industry tries to categorize both a person making a tape cassette for a friend and someone mass reproducing and selling that same tape as "pirates." They argue that whether or not the person reproducing copyrighted material is profiting is irrelevant because in both cases the technology enables a third party to consume copyrighted material without paying copyright owners. However, if it were not for the platform given to record labels by news outlets about the illegality of file sharing, the discourse over whether file sharing is in fact illegal may be different structurally.

The problem is that the idea of piracy has become naturalized. While it is more prevalent in media discussions, legislative chambers, and courtrooms than among academics, the view that digital music on the Internet results in "piracy" is prevalent in American society. There are also a lot of musicians and popular music studies scholars who believe in the necessity of maintaining copyright legislation to protect "the economic interests of the author and publisher" and who believe that the "author's right systems emphasize the spiritual connection between an author and his work" (S. Newman 1997, 22). Even authors who are critical of an overly protective copyright law foreground their argument based on ideological neoclassical economics arguments. [14] These scholars emphasize the wage system associated with music production and reify the position of the musician as an artist and owner, a position that I address at length. Since the recording industry profits from the sale of music, the industry constructs an argument claiming that circumventing the purchase of music is equivalent to stealing music. Certain behaviors are grouped under the term *piracy* and these behaviors are criminalized; the criminalization is justified by appeals to the victimization of copyright holders.

Digital technology, the recording industry argues, creates the means through which pirates can circumvent the industry's retail and distribution chains. Since music is a nonrivalrous consumer good (it can be used over and over again by many different consumers without being used up), the recording industry has argued that the purchase and digitization of one album can circumvent the need for anyone else to purchase that album (see http://www.riaa.com). In effect, digital music creates an environment where there is no longer a reason for people to pay for music since digital music disconnects access to music from the purchase of it. For the recording industry, the free distribution of digital music via the Internet is piracy because file sharing allows consumers to circumvent the established economic structure that the major record labels depend on to generate profit. The recording industry argued that the free distribution of music online made it very difficult for the major record labels to survive because they cannot compete with free music. If music fans downloaded their favorite music free, the RIAA asserts, then recording artists would have no source of income. The RIAA went as far as to claim that "employment at the major U.S. music companies has declined by thousands of workers, and artist rosters have been significantly cut back" because of file sharing (RIAA 2012). Additionally, the recording industry contends that without any income from music, those same musicians have no incentive to write and record new music; for instance, former U.S. attorney general John Ashcroft, in an interview supporting the RIAA's efforts to fight piracy, claimed that "we need to protect the creative works of the people in this country" (Butler 2004). Finally, the recording industry argues that recorded music as we know it would cease to exist without copyright because there would be neither incentive for musicians nor revenue for the record labels to record new music. While some recording artists may support the industry's position, the acceptance is far from universal.

"Piracy" is not written into copyright law, but rather it exists as a rhetorical statement about counterfeit goods. The part of copyright law that is conflated with piracy states that copyright infringement is the unauthorized *commercial* reproduction of copyrighted material, but piracy is the informal name given to such enterprises that produce large-scale counterfeit goods. Counterfeiting operations that mass-produce CDs, tapes, and vinyl albums for sale without mechanical licenses have been labeled pirate operations. Jessica Litman (2006) and Lawrence Lessig (2004) both highlight the commercial nature of piracy. In effect, the recording industry has created a narrative in an attempt to construct a framework for reinterpreting the law; James Boyle calls this the "Internet threat" (2008). By claiming that file sharing is piracy, the industry has made the argument that copyright infringement occurs in the refusal to pay, not just the unauthorized sale. While this book is not specifically about copyright and piracy, it is important to recognize the slippage in discourse that occurred during the digital transformation. "Today, the term 'piracy' seems to describe *any* unlicensed activity" (Litman 2006, 85). Two concepts in Litman's description are important; she uses the word "seems" to describe piracy because the law has not changed the definition of infringement, and the phrase "*any* unlicensed activity" points to the fact that there are unlicensed activities that are legal, per se. I argue that the recording industry has persuaded the public to believe that the sharing/giving of music is piracy through the piracy panic narrative even though neither the law nor tradition supports that position.

The piracy panic narrative gave major record labels an opportunity to seize public attention and legal support for the position that any circumvention of paying for music is an act of property theft. File sharing became stealing. Music listeners became pirates. The recording industry made this conscious shift before the conception of Napster and P2P file sharing programs.

TRANSMEDIATIONS HAVE CONSEQUENCES

Changes to law and technology affect the way the public creates and shares culture. This means that shifts in the music industry have wider implications than just for the production and consumption of music. First, changes in the distribution system facilitated increased oligopolistic control. Fewer, larger corporations now dominate the music industry—a trend that mirrors changes in other sectors of the economy. This has implications for "public culture," a conception that culture is created from the commons and cannot be solely attributable to one person. However, when large corporations control access to culture through copyrights, they aim to lockdown the public's access to cultural creativity. Copyrights construct boundaries around culture that define ownership, and record labels protect those boundaries. It means that independent artists are less able to compete with artists on the major record labels because they need access to copyright licenses at times to create new material. It also results in the employment of fewer recording artists as fewer labels require fewer artists. This leaves music listeners with fewer options.

Second, as the recording industry influences modifications to copyright law and Internet policy to stop online "piracy," it restricts legitimate uses of technology. Since the early 1990s, the recording industry has used the U.S. Congress and the judicial system to assert greater control over the media that provide content. In the copyright wars, the copyright industries implicate technology as inherently freeloading on their property. As a result, the legal system closed file sharing websites that transmit content in an unambiguously legal manor. "Information wants to be free" was a rallying cry of the cybernetic revolution, but in practice, the Internet has commodified information and knowledge in ways that were previously inconceivable. For instance, in 1998, Congress passed the Digital Millennium Copyright Act (DMCA), which fundamentally changed how copyright operates. Where copyright has always allowed the public certain rights (known as fair use), the DMCA blocks those allowances with digital rights management (DRM). Media encrypted with DRM do not permit users the freedom to do whatever they want with it. The simple way around DRM would be to break the encryption. However, the DMCA makes it illegal to circumvent DRM. For example, copyright law permits library-goers to copy up to 10 percent of a library book, but if the book is digital and encrypted with DRM, then the library-goer can only copy what the publisher permits instead of what the law permits. These legal and technological changes have a profound effect on public culture.

Third, changes in the industry are changing the content of music itself. From the sonic compression of MP3 files[15] to the increasing reliance on the "single" format, the digital transformation changes music creation and production. MP3 files are digital sound compressions that represent the actual recording, and as a result, these files do not have the same sound quality as uncompressed sound files. Producers now record music to sound optimal on MP3s. Additionally, since more people buy digital singles today, musicians, record label staff, and producers create more music that is capable of standing on its own as a single. With shifts in the media, there are parallel shifts in production as musicians begin to record music that works under the new media regime.

Finally, the digital transformation changes what it means to consume media. Whereas hard media allowed people to hoard music, digital media shifted toward a pay-per-use model. For music, this means that people are beginning to quit purchasing media and instead subscribe to services online. When those consumers no longer pay for their subscriptions, they lose access to the music to which they have been listening. This shift affects other media beyond music; DVRs are illustrative of changes through which people can consume recorded television. For instance, VCRs allowed people to tape shows and replay them anywhere, but DVRs only allow users to watch shows on their machines; again, if they leave their cable service, they lose any recorded shows. Music and movies remain the most dramatic shift here because people often pride themselves in their collections; digital media are eliminating the need for a collection.

While books and articles from the academic (Burkart and McCourt 2006; David 2010; McLeod 2005; Park 2007) to the journalistic (Knopper 2009) have thoroughly described this transition, they all start from the standpoint that the Internet and digital technologies surprised the recording industry. For instance, after predicting the "death of the major record labels," Matthew David points to a decline of the "gatekeeper function" that undermines the labels' monopoly (2010, 7). David projects that the record labels are worse off today because of the digital transformation, and he goes as far as to claim that the late 1990s were the "golden age for the recording industry" (2010, 33); however, according to my research, the industry appears to be stronger financially today than before. Patrick Burkart and Tom McCourt claim that it is "inarguable . . . that the industry is deeply unsettled" and that it must "either change or face obsolescence" (2006, 18). The idea that the recording industry is unsettled is problematic enough, but to contend that it must change simply misses the fact that the recording industry has been the catalyst of change during the digital transformation. By arguing that the digital transformation was a reaction to Napster, these accounts miss the fact that the recording industry was initiating changes to the means of production/distribution and the legal apparatus that creates this system as early as 1992, long before the development of Napster.

Rather than taking the RIAA's self-description of victimhood at the hands of "pirates" as descriptive of their experience with this digital transformation, my project examines the past two decades in an attempt to understand the material changes within the recording industry. I contend in this book that shifts in the law, technology, and the commodity were initiated through efforts by major record labels and the RIAA to transition to a more economically efficient means of production.

THE BOOK GOING FORWARD

To stop the supposed decline of recorded music, major record labels and the RIAA have attempted to change the structure within the music industry in four ways: (1) the RIAA and major record labels changed the nature of the recorded music commodity; (2) the RIAA adjusted the role of the state in regulating the recording industry; (3) major record labels altered their relationship with recording artists (their labor) through record contracts; and (4) the RIAA and major record labels used surveillance to encourage music consumers to consume music in the industry's desired manner. Each part of this book analyzes one of the ways that the recording industry has changed the political-economic structure of the broader music industry.

Part I looks at how the commodity of recorded music has changed during the digital transformation. More broadly, it explores the ways the recording industry has transitioned to digital music. I explain that pronouncements of the industry's struggles and imminent decline were not only exaggerated but also disingenuous. Rather, as Part I demonstrates, the political economy of the recording industry is much more complex than suggested simply by the sales figures that the industry reports through the RIAA. To make this argument, I conduct a close reading of data from both the International Federation of Phonographic Industry's (IFPI)[16] *Recording Industry in Numbers* and Nielsen SoundScan's annual reports. This part demonstrates that the major record labels have actively changed the recorded commodity during the digital transformation to expand the means of consumption. Additionally, I show that a change in the recorded medium brings about changes in the form and content of the commodity.

Part II analyzes the way that the state's role in the recording industry evolved during the digital transformation. I do this through an examination of several policies passed by the U.S. government over the past two decades. Two main areas of state media policy affect the recording industry: copyright law and Federal Communications Commission (FCC) regulations. The recording industry did not passively observe the digital transformation while waiting for the courts and legislatures to interpret and design laws that accommodated digital media; rather, Part II explores the concrete policies that the RIAA advanced in order to use the state to reinforce the major record labels' dominant position in the music industry. These laws not only address copyright and distribution but also change what it means to "own" music. As a result of this legislation, many music consumers have shifted from music owners to music users. Part II analyzes how different groups are involved in the legislative process. Furthermore, this part demonstrates that since Internet users have established consumption habits on the Internet, new policies, such as net neutrality, do little to change their patterns of consumption.

Part III examines how the relationship between the major record labels and recording artists (their labor) has changed. As file sharing hit its peak from 2000 to 2003, the RIAA insisted major recording artists were victims of piracy. In this way, the recording industry attempted to position file sharing by music fans as a direct threat to musicians' livelihoods. Part III begins by exploring the way musicians function as labor in the recording industry. Through a close reading of the Senate Judiciary Committee hearing on Napster in 2000, I elucidate how the recording industry constructed musicians as victims of property theft. While the recording industry argued that file sharers were victimizing recording artists, I show that the major record labels seized the opportunity to change the structure of recording contracts by switching to "360 deals."[17]

Part IV examines the ways that major record labels and the RIAA used surveillance to encourage music consumers to consume in the recording industry's desired manner. As music fans turned to free music online, major record labels argued that having consumers pay for music was vital for the industry's survival. By suing file sharers for allegedly violating copyright law, the RIAA created a system where, for many people, the costs outweighed the benefits of using P2P programs. At the same time, Part IV explores the way major record labels used market surveillance on P2P programs to understand the consumption patterns of music fans. Part IV demonstrates that this dual surveillance was an active attempt on the part of the recording industry to direct the digital transformation of the recorded music commodity.

Part I

Transformations in the Recording Industry

Chapter One

Recording Industry in Transition

Between 2000 and 2010, there were few reports about the strength of the recording industry. Instead, most news articles pointed to the devastating effects that digital distribution, and more commonly "piracy," has had on the major record labels and their artists. For instance, a *Washington Post* article discussing the sale of Warner Music Group says that the sale was the result of "an industry struggling to boost digital revenue to offset the impact of falling CD sales and piracy" ("Access Industries to Buy Warner Music" 2011). Similarly, a *CNET* article pronounces that "digital music and music players, as well as the rise of illegal file sharing, helped to hasten the demise of the CD as the main music distribution format" and contributed to the closing of a CD manufacturing plant in New Jersey (Sandoval 2011). These articles neatly pick up the piracy panic narrative and assert that declining CD sales are the consequence of file sharing. However, the causality between digital music piracy and declines in CD sales that both articles point to is not as evident as they assume.

The reality of the digital transition cannot be fully grasped without considering the transition of previous media formats. The political economy of the recording industry is much more complex than the sales figures reported by the RIAA, a reality that I reveal through a close reading of data from the International Federation of the Phonographic Industry's (IFPI) *Recording Industry in Numbers* and Nielsen's SoundScan reports, which have tracked sales over the past two decades. While CD sales declined beginning around 2000, this decline is the product of a routine transition between recording media commodities. I challenge the recording industry's orthodoxy and assert that the recording industry was not a passive actor in the digital transition of the music commodity, but rather the recording industry actively changed the music commodity from CDs to digital files.

Publicly, the recording industry claims that revenues began to decline sharply in 2000 due to file sharing; however, in *Recording Industry in Numbers* (*RIN*),[1] the industry is more candid about the impact of other factors. The IFPI acknowledges that there is reason to believe that "the recent fall in CD sales [is attributable] to a maturation of the 'CD-replacement cycle' in the largest markets, whereby consumers have repurchased albums on CD that they had previously bought on cassette or LP. The amount of direct replacement of titles purchased on CD has never been researched and is unknown" (IFPI 2002, 8). In the North American market, the IFPI also claims that the CD's decline was linked to "uncertainty about the economy" and "increased competition from a revived gaming industry and DVD video" (IFPI 2002, 24). In the introduction to the 2005 edition of *RIN*, IFPI's market research director, Keith Jopling, is quite candid:

> The music business is emerging from a period frequently described in recent years as a "crisis." The year 2004 was the industry's best year-on-year performance for five years—and while tough times lay ahead, the industry has clearly turned a corner. . . . Demand for music, in all forms, is higher than it has ever been before. And entertainment sectors related to music, or that use music, are growing. But there are big challenges for the music market too. Internet piracy is the greatest of them, but it is by no means the only obstacle to growth. Consumers are spending more on other, newer entertainment products such as DVD and games. They are also spending more on mobile phones, internet and other "subscriptions." (IFPI 2005, 3)

So while the IFPI and the RIAA were publicly blaming file sharing as the culprit for the decline of CDs, the IFPI states candidly that the "free market," recession, and a buying cycle contributed to the format's decline. This quote reveals the complexity of the recording industry's challenges with digital media; however, Jopling ensures that industry readers keep the piracy panic narrative at the forefront of all discussions about digital music.

Additionally, studies by economists have been inconclusive, at best, as to whether file sharing affects music sales (Connolly and Krueger 2005), and a growing number of economic studies have demonstrated that file sharing had no effect on music purchases (Andersen and Frenz 2010; Oberholzer-Gee and Strumpf 2007). But the major record labels continue to point to piracy as the reason for revenue declines. The recording industry equates downloading music with theft by stating that "digital music theft has been a major factor behind the overall global market decline" (RIAA 2012). To demonstrate its point, the RIAA and the IFPI provide the information displayed in figure 1.1 on revenue.

At first glance, figure 1.1 appears to show that U.S. music revenue drops off precipitously beginning in 2000. This graph presents an idea that, in 2000, industry revenue already lost $200 million from a 1999 record high.

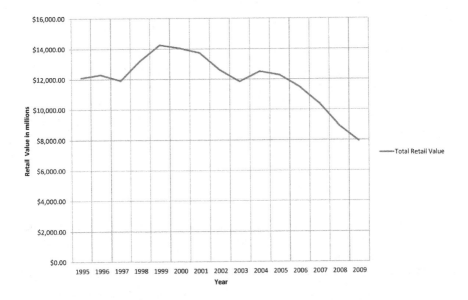

Figure 1.1. Total Retail Value of Music in the United States (1995–2009) according to the IFPI

Yet a number of questions remain unanswered. What caused this decline? How is this revenue calculated? Why were revenues so much higher in 1999 than in previous years? How accurate are these data? The industry maintains that file sharing caused this drop in revenue. However, these numbers are based on shipments (the number of units shipped to a store) rather than actual units sold. Furthermore, these data assume that each unit is sold for the average price of a CD each year; however, CDs rarely sell for the average price, and this further inflates the discrepancies between actual sales and shipments. What we see here is a partial story, and this partial story has dire consequences on the rhetorical playing field around copyright because the recording industry uses it to create new policies to combat "pirates." In this chapter, I take a critical look at these data to understand where they come from, how they are produced, and what the effect is on the recording industry.

For more than seven decades, the recording industry has been the dominant cultural producer within the music industry, but even within the recording industry, the primary commodity has shifted at different times. Consumers fetishize the recorded music commodity by connecting the existence of music to its recorded form, but music does not have to be recorded to be music, or even to be a commodity. The recording industry's dominance was the result of the industrialization of music when a reliance on the distribution of a physical recording became the dominant means of production. Theodor Adorno reasons that "the transition from artisanal to industrial production transforms not only the technology of distribution but also that which is distributed" (2002a, 271). Adorno considers how music production changed from sheet music and home performance to the recording industry's reliance on vinyl records and the mechanical reproduction of sound. Today, the transition from industrial to digital production that changes the distribution, production, and consumption of the music commodity renders it fundamentally different from its previous form.

RECORDING INDUSTRY IN TRANSITION

A definite transition took place in the recording industry between 1995 and 2010. The music commodity changed from a physical medium that exists primarily in the auditory dimension to a virtual commodity that increasingly blends music with images and text. The term I use for this change is *transmediation*—the change from one medium to the next. Transmediation tends toward a synthesized and synchronized format that crosses conventional boundaries between music, video, and text.[2] In the past, the control over the physical production and distribution networks gave a small number of major record labels an oligopoly over the recording industry and the broader music industry.[3] However, the recent transmediation did not reduce the power of the major record label oligopoly—in fact, the oligopoly consolidated power as it went from six major labels to three (Arditi 2014). The physical medium of the CD is of less importance to labels, but the move away from the CD as the main physical medium of the recording industry is just one moment in the history of recording media. Transmediation has been a part of the recording industry's history since gramophone records replaced phonograph cylinders as the main medium of the recording industry. Since the medium of recorded content has regularly changed, it is important to delineate the difference between the medium and the content; the recording industry sells content on a medium, and the major record labels are only nominally concerned with the medium. During the recent transition, and much like previous transitions, the recording industry found new ways to create revenue through new commodities, while they adapted new business models to new technologies.

While record labels continue to produce physical media, the commodity that they have always sold is the content of recorded music, not the medium. In other words, record labels sell songs not CDs; more specifically, labels sell the copyright licenses to songs. At different points in the history of the recording industry, recorded music used different media formats. The first media of the recording industry were phonograph cylinders (played on the phonograph) and gramophone discs (played on the gramophone), as the Columbia Graphophone Company and the Victor Talking Machine Company competed against each other by producing music to play on their respective machines (Chapple and Garofalo 1977; Frith 2006, 233). Today, the main media formats are digital music files, but the fact remains that labels claim to own and sell content and copyrights, not the physical medium. No one declares that the decline of phonograph records was a marker of the end of the recording industry. Neither do labels claim that the decline of tapes signaled the death of the recording industry. The recording industry does not make these assertions because new media always replace old media. Since the recording industry is not committed to a particular media format, the best assessment of the strength of the recording industry is to measure the total volume of sales across media formats rather than measuring any particular medium.

At different times the recording industry has pushed for changes in medium (i.e., changing from tape cassettes to CDs), while fighting transitions at other moments. The recording industry has resisted the deployment of new technologies when a medium lends itself to reproduction and, generally, has accepted technologies that are more secure. During the digital transition, the recording industry publicly resisted the transition from CDs to MP3s, while the major record labels quietly developed secure digital technologies. This resistance to the MP3 is evidenced by the recording industry's formation of the Secure Digital Music Initiative (SDMI) to create "secure" music as opposed to the "open" format of MP3s (Reece and Jeffrey 1998). Since SDMI developed before Napster launched, the recording industry was proactively developing a means to mass-produce digital files that would limit the sharing of music. Therefore, the digital transformation was no different from previous transitions because the industry attempted to manage the transition through the creation of secure media.

During the digital transformation of music, the main medium of music significantly changed. In 1995, the recording industry's primary commodity was the CD. Nielsen SoundScan data show that the recording industry sold 616.3 million albums in the United States in 1995, which includes 409.5 million CDs and 205.8 million cassettes (Nielsen 1997). However, the change to CDs was only the most recent music format transmediation. Compact disc sales in the United States rose to two-thirds of the market in the decade following their commercial release. The benefits for music consumers of purchasing CDs were the relative durability of CDs over tape cassettes and the portability of CDs over LPs.

However, the transition from cassettes to CDs was not driven by consumer demand, but rather the recording industry intentionally initiated this transition. There was a recording industry policy in which major record labels began to refuse to purchase unsold tape cassettes back from retail outlets in the late 1980s to force record stores to sell CDs (McLeod 2005).

This was a critical point in the rise of the CD because music retail stores require the safety net of record labels buying back unsold albums to protect the retailers against the uncertainty of consumer demand. Since the sales of any given album/single cannot be known in advance, retailers have no idea how many albums to stock. Because retailers could no longer rely on the major record labels to buy back unpurchased cassettes, they were forced to sell CDs and consumers were forced to buy them. Figure 1.2 shows the IFPI's global data on music shipments between 1973 and 2007. This graph provides a useful visual demonstration of the way that new media replace old media; new media experience a sharp increase in sales followed by a steady decline. There is a striking pattern in figure 1.2 in that new media follow a similar life cycle, but also each new medium peaks about 1.5 times higher than the previous medium. While figure 1.2 exhibits the existence of a reliable pattern for new media in the music industry, it also points to a steady expansion of consumption as the volume of shipments increases.

Transformations of the recorded medium in the recording industry happen periodically out of a general process through which record labels try to get consumers to repurchase music that they already own. However, the recording industry no longer controls the development of new media on which to play music—that is, the recording industry produces software, whereas consumer electronics manufacturers produce hardware. There is a symbiotic relationship between record labels and consumer electronics manufacturers: consumer electronics manufacturers need the content owned by the recording

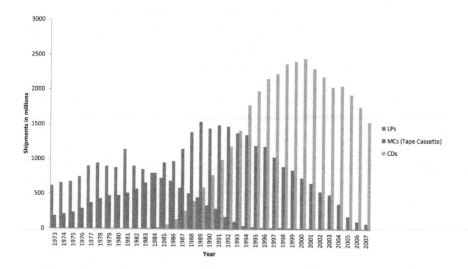

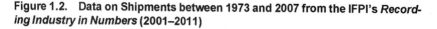

Figure 1.2.   Data on Shipments between 1973 and 2007 from the IFPI's *Recording Industry in Numbers* (2001–2011)

industry to play on their machines, and the record labels need new machines to resell their music in new formats to music consumers.[4] After the major record labels became established, there was no longer a reason for hardware companies to produce software to use with their equipment. Rather, hardware companies began to seek licenses from record labels to gain the right to play music on their machines; generally, record labels have been receptive to this licensing because it gives them a medium through which to sell music. Record labels are not always opposed to the idea of technological change but rather embrace it, particularly if it means music fans repurchasing their catalog in a new format.

While the relationship between consumer electronics manufacturers and record labels is symbiotic, at the same time, it has always been contentious. During the early days of the recording industry (1890s–1950s), competition occurred between record labels over record formats and player formats (Chapple and Garofalo 1977). Since the content was copyrighted by labels, this meant that owning a specific record player only allowed consumers to play records from the company that produced that specific machine. There were fierce format competitions, similar to the recent format duel between HD DVD (Toshiba) and Blu-ray (Sony). In the 1950s, the major record labels recognized that this competition among formats was detrimental to consumption, and they decided to work together by forming the Recording Industry Association of America (RIAA) in 1952. The RIAA was created to regulate the equalization curve of records—a formula that changes the playback of records (Hoglund 2011). While various record players could interchangeably play records from other record labels, the sound was "muddy" (Hoglund 2011) when the label did not match the player. This format truce lasted until the development of stereo records in 1957 by EMI (Chapple and Garofalo 1977). Following the adoption of stereo by the major record labels, music fans went to music stores to repurchase albums they already owned in mono to hear their music in stereo, and "overall record sales were boosted as a result. To phase mono out and make an extra profit the majors raised the price of mono records to equal the stereo price" (Chapple and Garofalo 1977, 53). When new formats become available, the recording industry generally pressures retailers and consumers to accept the new format in order to benefit from the repurchase of music. This process is evident in the recording industry's sales numbers at moments of transmediation.

Taking a closer look at the sales figures for CDs, it is evident that there has been a transition of recorded medium occurring over the past decade and a half. In 2010, the recording industry showed a tremendous shift from 1995 as sales of CDs have plummeted since peaking in 2000 (according to the IFPI) or 2001 (according to Nielsen); CD sales declined by roughly half between 2001 and 2008 according to both Nielsen SoundScan and the IFPI (see figure 1.3). The difference between Nielsen SoundScan data and IFPI/

RIAA shipment data is significant, too, because it points to some of the problems with accepting the recording industry's piracy panic narrative. There has to be a reason why the RIAA showed a decline in 2001 and Nielsen did not show a decline until 2002. For now, my point is that CD sales have in fact declined—this is indisputable. But why does the recording industry point to the decline of the CD format as the harbinger of the end of the music industry? The decline of CDs is not representative of the industry's longevity if a new media format takes the old media's place.

During the transition to digital music, overall music sales in the United States reached new highs. According to Nielsen SoundScan, overall music sales equalled over 1.5 billion units in the United States in 2010. "Overall music sales" is the phrase Nielsen SoundScan uses for all music sold in a given year; this includes CDs, LPs, tape cassettes, digital singles, digital albums, and so on, though it still does not account for revenues from performance rights, synchronization, or publishing rights. Since the recording industry sells more than just CDs, it is important to look at overall music sales because it gives a better idea of the extent to which people consume music. Nielsen only began tracking overall music sales in 2003 as a unique category,[5] but over that short time period sales have more than doubled from 687 million (see figure 1.4). Data in figure 1.4 are different from data in figure 1.3 in that figure 1.3 shows only CD sales whereas figure 1.4 shows sales for nearly *all* music media sold. The recording industry has managed to increase the sheer volume of sales by nearly 100 percent in the United States over the ten years since the peak of the CD.

These data are staggering, but according to the recording industry, major labels are struggling to survive in the face of rampant piracy (RIAA 2012). In 2010 *RIN* stressed that piracy is a major problem for the recording industry (IFPI 2009) without attempting to reconcile the fact that music sales were at an all-time high. The problem with the recording industry's claims about declining sales numbers also rests in the acquiescence of the news media in furthering the piracy panic narrative when the RIAA and the IFPI make claims about the recording industry's weakness. Part of the effectiveness of the piracy panic narrative is the collaboration between the recording industry

Year	1995	1996	1997	1998	1999	2000	2001	2002	2003	2004	2005	2006	2007
RIAA Shipments	722.9	778.9	752.9	846.1	933.8	942.5	881.9	803.3	746	767	705.8	619.8	511.1
Nielsen Sales	409.5	448.4	504.6	578.3	648.1	706.3	712	649.5	635.8	651.1	599	553	449

Figure 1.3.   CD Shipments (RIAA) and CD Sales (Nielsen SoundScan), 1995–2008, in the United States (in millions)

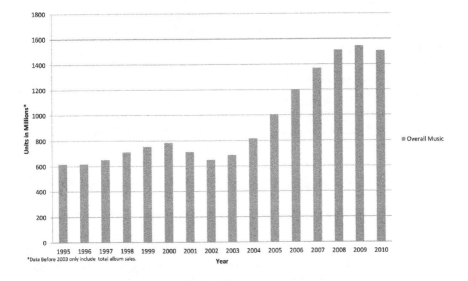

Figure 1.4. Overall Music Sales in the United States according to Nielsen SoundScan (1995–2010)

and the news industry on copyright issues. Both news corporations and record label corporations are owned by larger media conglomerations; this is an issue that media scholars often discuss in terms of the problems with media conglomerates (Baker 2007; Herman and Chomsky 2002; McChesney 2004). Media organizations are sometimes called "copyright industries" because much of their product is derived from copyright laws. When newspapers and news shows cover issues of piracy, they are more than willing to substitute journalism with press releases from the RIAA.

The decline of the album changes music (see figure 1.5), but again this should neither come as a surprise nor be lamented by audiophiles. Albums are only the product of a business model that worked at a certain time for the recording industry, and we should not expect them to remain as the primary organizational tool of recorded music in the future. While album sales were at 785.1 million units in 2000, they dropped to 443.1 million in 2010 (Nielsen 2001; Nielsen 2011). This means that the demand for and sale of albums has drastically changed over the decade. Some critics have argued that this amounts to the destruction of an art form (Ulaby 2003), and it is clear that the feature length album is in decline. Technically defined as "a recording containing usually eight or more individual songs or 'cuts,' totaling 30 or more minutes of playing time" (Hull, Hutchison, and Strasser 2011, 318), the album is a particular recording format that enables record labels to sell multiple songs on a single medium. Following Adorno's (2002a) contention that technology helps to determine the content that gets produced and distributed, an art form, like the album, that corresponds to a particular medium, like the CD, is partially limited in design by the technology available to create the commodity. Whether music produced by major record labels is sold as an album or a single is determined by business decisions. Therefore, albums are a product of the recording industry's pursuit of profits at a particular time in history, not the demands of consumers. Part of the reason that the recording industry is not producing/selling as many albums now is that it has become more efficient for the recording industry to create profit in other ways. Constructing the album as art assumes an artistic autonomy that never existed—it obscures the commodity character of the album and its profit function. While the distribution and media format of music has real effects on the final product in the creative process, it is important not to reify the corresponding formats of music because they are products of capitalist production processes.

In contrast to the decline in album sales, there has been a historic rise in the sale of singles, specifically digital singles. According to the IFPI, single sales were at 102 million units in the United States in 1995 and declined to 8.4 million units in 2002 (IFPI 1996; IFPI 2003). However, following the creation of online music retailers, single sales rose to 1,160 million units by 2009 (IFPI 2010).

The rate at which the digital single increased between 2003 and 2009 was unprecedented, but the single as a dominant media format is not new. Figure

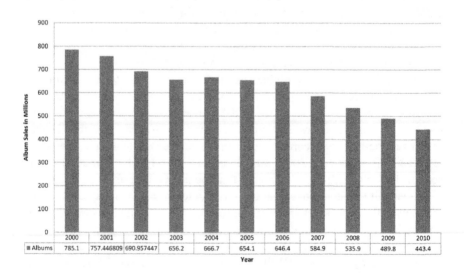

	2000	2001	2002	2003	2004	2005	2006	2007	2008	2009	2010
Albums	785.1	757.446809	690.957447	656.2	666.7	654.1	646.4	584.9	535.9	489.8	443.4

Year

Figure 1.5.   Album Sales in the United States according to Nielsen SoundScan (2000–2010)

1.6 gives a long-range picture of global singles sales; this demonstrates that the sale of singles has experienced peaks and valleys through different media. Until the advent of the 78-rpm record in the 1950s, music was primarily available in the singles format because each side of a record could only support one song (Hull, Hutchison, and Strasser 2011); this was also a technical limitation on the length of a song because a song could not exceed the available space on one side of a 78, which was about three and a half minutes. Sales of singles peaked in 1983, and despite enjoying a slight resurgence in the mid-1990s (see figure 1.6), declined throughout the 1980s and 1990s.

Overall, the decline of the single between 1983 and 2003 was a product of the recording industry's own doing because it was more efficient to sell multiple songs on a single medium for a higher price. While the IFPI tries to connect the single's performance to demand (IFPI 2002, 8), my research asserts that the decline of singles was linked to the cost of producing a single versus the cost of producing an album. A CD costs roughly one to two dollars to produce (Park 2007) regardless of how much content is on the CD. In terms of CD singles, "the single cost too much" (Knopper 2009, 106) because even though the physical CD cost one or two dollars, a single can only be sold for a fraction of the price of an album. Since the "suggested retail list price in the United States has for some time been $6.98 for singles and $16.98 to $18.98 for albums on a CD" (Krasilovsky et al. 2007, 20), the revenue from a CD sale increases by ten to twelve dollars for an entire album over a single.[6] Therefore, decisions on whether or not a major record label

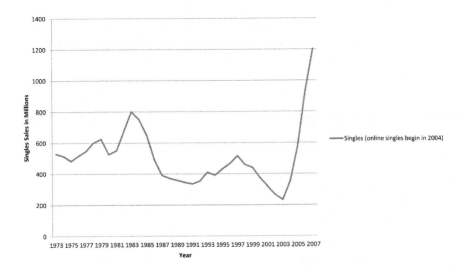

Figure 1.6.   Global Single Sales according to the IFPI (1973–2007)

will sell full-length albums or singles are determined by the projection of the profit that they can make from a sale. The use of the single as the primary commodity of the recording industry is not new, and it is unlikely that its status will remain the same in the future because new media will provide different ways to distribute music.

The transition from CDs to MP3s is not altogether different from the transition from LPs to tape cassettes in the early 1980s. When tape cassettes became widely available, figure 1.6 shows that singles sales increased significantly. However, this transition to singles on tape cassettes was not met with the same epitaphs about the death of the album that occurred with the rise of digital music. The type of media format that the recording industry sells has changed over time, but the recording industry has led these shifts.

SHIFTS IN THE COMMODITY FORM

Shifts in recorded media also affect the content and form of the music. In the case of the digital transformation of the recorded music commodity, the confluence of technology and distribution affects the realm of content and form because digital music enables the proliferation of singles.

The development of the recording industry created a mediated system where music fans no longer had direct access to the creation and performance of music as music; recordings substitute for live performance.[7] Recording technologies and music corporations mediate music consumption. A key to the mediation of music is its distribution—"the processes whereby music comes to be allocated among different audiences" (Negus 1999, 95–96). Distribution mediates through people and technologies, that is, industry gatekeepers. These gatekeepers (radio, retailers, and artists and repertoire[8] staff) affect the content of music before the first consumer has an opportunity to make listening decisions. Large retailers have become a force by determining what they will and will not place on their shelves and make available to consumers. For instance, Walmart will only sell music that it deems "family friendly" (Klein 2002); consequently, recording artists internalize Walmart's censorship policies when they record new music. The system of distribution has as much an effect on music as the musicians that create the music because distribution defines the parameters of what retailers can accept under a given media regime. Distribution is the way through which music fans have access to new music. If music fans cannot get to a new recording, it essentially does not exist. As distribution systems shift, the access that music fans have to new music shifts as well.

Digital distribution of music causes the disintermediation of the music commodity. Disintermediation is the removing of intermediaries in the process of distributing media. For the recording industry, this means that music does not go from the recording studio to a manufacturing plant to a distributor to a retail outlet to the consumer, but rather goes directly from the recording studio to the consumer by way of an online retailer. Disintermediation greatly reduces costs for the record label as intermediaries are eliminated from the process; however, musicians and songwriters still get the same royalty rates as in an intermediated distribution system.

It is best to view the shifts in distribution and medium of the music commodity as the product of the technology available to and accepted by the recording industry. The relationship between medium and recording limitations dates back to the gramophone. For instance, Adorno criticized that the "only thing that can characterize gramophone music is the inevitable brevity dictated by the size of the shellac plate" (2002b, 278). If the physical medium can hold only five minutes' worth of music, songs produced in that format will be limited to five minutes in length. If a medium can hold sixty minutes of music, then record labels will develop a way to fill that format with as much music as possible. Because audiences were already used to songs that were the length of a "side," songs have generally adhered to those length restrictions—about three and a half minutes. At different moments, the recording industry has adopted different forms of recordings that integrate with different media formats. For instance, concept albums are longer compositions unified around a similar theme where each song on an album becomes a movement. These longer thematic albums were the result of there being more space available on an LP to record more music. In turn, the concept album was based on the idea of creating one overall artistic work in the form of multiple "sides." There is a direct relationship between the medium and the commodity; as changes occur in the music commodity's medium, the form (i.e., single or album) of the commodity is changed by the recording industry to better market the music commodity.

When there is a shift in the recorded commodity form, it is generally driven by the record labels themselves. Record labels change the form of the music commodity at moments of transition to enable larger profits. The production and distribution costs of producing one unit of any given medium (LP, tape, CD) are virtually the same whether the physical unit contains five or sixty minutes of music, and, given that a label can charge more money for a full-length album, profit margins are larger for full-length albums in physical formats than for singles. However, part of the problem with full-length albums is that recording artists are forced to include songs that are potentially not as "good" as others are because they need to fill the format, or as Steve Knopper describes it, "the record business had boiled down much of the business to a simple formula: 2 good songs + 10 or 12 mediocre songs = 1 $15 CD, meaning billions of dollars in overall sales" (2009, 106). While this holds true for the physical CD, digital music eliminates the need to produce the ten to twelve mediocre songs, or at the very least, it allows record labels to assess the commercial viability of the additional songs as singles. In this way, paying for music on a per song basis means that record labels only have to produce what they think will sell without a need for filler to make an entire album.

While music fans and musicians are apprehensive about changes in the commodity, the commodity itself is only the product of the demands of the recording industry. When the recording industry switched its main commodity from the single to the full-length album, the major record labels calculated that they could sell twelve songs for more than one song on a physical commodity that cost the same to manufacture and distribute regardless of how much content is on the physical medium (Knopper 2009, 104–107). Digital singles flip this logic on its head once again, as it costs virtually nothing to distribute digital music, but every song recorded costs more in studio time and production labor. Instead of producing one album with twelve songs and only two "good" cuts (Knopper 2009, 106), a record label can pay to produce fewer songs that will be more likely to sell on their own. I speculate that in the future, record labels can have their recording artists go into the studio and record only those songs that both the musicians and the label feel confident will generate attention from consumers. In fact, Justin Bieber records singles not associated with albums, releases them regularly, and later creates compilation albums of those singles. This eliminates wasted time in the studio when recording artists are pressured to produce results from thin air. Digital singles make sense for the recording industry because there is no longer a financial need to produce filler on an album.

ALBUM/CD REPLACEMENT CYCLE

Each new media format creates new profits for the major record labels as music fans replace their album collections in new media formats, creating a rise in overall album sales driven by an increase in catalog sales. As Kembrew McLeod explains, the policy of pushing retailers and consumers toward the CD "generated higher profits and new sales as fans began replacing their vinyl collections with CDs at inflated prices" (2005, 526). David Park describes this process as the "CD replacement cycle" and argues that the CD "was premised on the assumption that baby boomers would be likely to replace their vinyl collection with CDs" (2007, 72), which I call the "album replacement cycle." When the recording industry goes through transitions, part of the increase in profits that labels experience is driven by consumers wanting to hear the music that they already own on their new music players. With each new media, record labels expect to see an increase in revenues driven by music consumers repurchasing albums they already own in their collection. It is important to understand the relationship between music sales volume and the purchasing habits of music consumers (especially with respect to the music that they already own in a different format) in order to understand the role that digital music plays in the decline of the CD.

Since there was no need for music fans to repurchase their music as digital music files, new digital media did not create a spike in music sales for the recording industry. By the late 1990s, music fans had already replaced the music that they owned on tape cassette and vinyl record over the previous decade. In the late 1990s and early 2000s, music fans began to replace CDs with MP3s to use on computers and portable MP3 players. One major difference exists between the transition to CDs from LPs and tapes and the transition from CDs to MP3s: music fans had no reason to purchase MP3s in order to replace their CD collections. Computer technology allows people to "rip" CDs onto computer hard drives by placing a CD in a CD-ROM drive and saving the music files. Since the CD rippers already own the CD, the personal copying of the CD is permitted under the Audio Home Recording Act (AHRA) of 1992[9] and *Sony Corp. of America v. Universal City Studios, Inc.* (464 U.S. 417, also known as the "Betamax Case"). While the copying of one's own music collection onto a computer interferes with a process of cash flow to the major record labels that they have been used to exploiting since the early days of the recording industry, this type of copying is legal.

A further reason for the end of the album replacement cycle was the lack of availability of MP3s for purchase. Music fans could easily rip music from CDs onto hard drives, but digital music retailers had small music libraries in the early years of MP3 files. In fact, the IFPI partially blames the early days of music file sharing on the lack of "legitimate" alternatives to peer-to-peer (P2P) file sharing (IFPI 2002). Since music was not available in MP3 format for consumers to play on their computers and MP3 players, they were turning to alternative ways to obtain their music.

However, P2P services provided the means for music fans to download music that was not otherwise available online. Lawrence Lessig counters the IFPI's claim by describing P2P services as a means to find deep catalog music that is difficult to find in brick-and-mortar stores; he further maintains that these services function more like a library (Lessig 2004, 68). Furthermore, some of these recordings found on P2P networks could be ones that music fans had owned in a previous format (i.e., vinyl) but did not replace on CD and are therefore unable to rip onto a hard drive. These alternative methods of finding music conflict with the recording industry's business strategy, but they do not violate copyright law. By contending that the downloading of music through P2P networks was due to a lack of "legitimate" alternatives (i.e., record-label-approved Internet stores), the IFPI negates these alternative and "legitimate" (i.e., legal) uses of P2P networks.

The recording industry has responded by attempting to develop media formats that are not reproducible by consumers and developing what it sees as a "legitimate" retail service from which to purchase music. While ripping music from CDs allowed music fans to use music in whatever way they wished, digital rights management (DRM) code implanted in digital files restricted the ways that music consumers can use the digital music they purchased (Gillespie 2007). For instance, digital ringtones became a popular fad for many cell phone users in the early 2000s, but users could not easily make ringtones out of regular MP3 files. In order to have a master ringtone (i.e., a ringtone of an actual recording), cell phone users had to purchase the ringtone separately because users could not transfer MP3s from a hard drive to a cell phone. In this way, someone who purchased a recording for his or her general music collection had to repurchase the same song if he or she wished to set the song as a ringtone.

The recording industry ensures that fewer non-encoded songs are available online by tethering music files to specific devices through DRM. As more closed[10] devices (such as smartphones and tablet computers) come into production, their proprietary software will execute only the coded files to work on that device. While ringtones and MP3s are distinct files that cannot be interchanged with each other on their corresponding devices, in the future, certain files may be used on phones, different files on MP3 players, and yet another file type for car stereos or a home theater system. Apple's iTunes already had files that could only be played on Apple's iPod MP3 players, and even though that particular DRM code is not being used, proprietary devices are becoming significantly more common. By restricting the transferability of digital music from one device to another, the recording industry managed to develop a system where music fans still purchase multiple copies of a recording, and the major record labels have been able to maintain their dominance.

Since the publication of Lessig's book *Free Culture* (2004), the recording industry has heavily invested in making their deep catalogs available on the Internet, and Lessig's original argument about P2P programs as libraries is less relevant because of industry-based transformations in the availability of catalog material. Following the rise of iTunes and other online retailers, record labels have cheaply added their deep catalogs to Internet databases. Record labels can rip master recordings onto large hard-drive farms at virtually no cost when compared to the expense of printing, distributing, and stocking deep catalog recordings at brick-and-mortar stores. The IFPI reports that at "the beginning of 2008 catalogues in the major online music services totaled 6 million songs—many times more than a traditional record store" (IFPI 2008, 4). With the expansion of online music catalogs, there is even less reason for record stores to stock older music. Furthermore, the major record labels aim to make their entire catalogs available online. In 2008, the IFPI "expected that over the next 3 to 5 years the vast majority of companies' catalogues will be available digitally" (IFPI 2008, 4). Following the expansion of online catalogs, the industry will try to convince consumers that they should purchase digital music to replace their personal physical collections.

The rerelease of the Beatles' catalog on iTunes is an example where the digital release of a deep catalog on the Internet resulted in fans repurchasing their personal collections. As the iTunes catalog grew after 2003, the Beatles' music was notably missing from the iTunes Store. However, on November 16, 2010, Apple released the entire Beatles catalog. Until that moment, the Beatles' music was available on MP3 only by ripping it from CDs or finding it online via P2P programs. *Billboard* identified two groups that would likely be drawn to iTunes for the Beatles' catalog: collectors looking to buy the box-set and a new generation of consumers (Bruno 2010b). Accompanied by an aggressive TV advertising campaign, the Beatles release on iTunes was received by the media as if no one had been able to listen to digital versions of Beatles songs until then. Individual songs were made available at the premium price of $1.29, individual albums were available for $12.99, and the entire catalog could be purchased for $149 (Ogg 2010). The release of the Beatles' catalog on iTunes spurred the sale of over two million individual songs and 440,000 albums in the week following the release (Serjeant 2010). At the same time, the release did not cost the Beatles, EMI (the Beatles' label), or iTunes to reproduce the music, aside from marketing. Since people are not necessarily looking for deep catalog music, "often consumers have to be driven to those titles by marketing" (IFPI 2008, 4). This was a key moment in the digital transformation of the music commodity because it marked the point when record labels developed a way to monetize and capitalize on the digital replacement of already-owned physical media.

The increasing availability of deep catalog music online will increase the ability of the major record labels to reinitiate an album replacement cycle for digital music. Despite the declining sales of CDs at the turn of the twenty-first century, due in part to the saturation of music consumers repurchasing their catalogs and the lack of a need to purchase digital music to replace personal CD catalogs, the availability and marketing of deep catalogs online created a new boom in sales. In fact, the IFPI even claims that catalog sales (titles over eighteen months old) "represented 39% of overall album sales in the US, up from 36% in 2001" and constituted 46 percent of all online sales (IFPI 2008, 4). Digital catalog sales indicate a triumph for the recording industry in its ability to replicate the album replacement cycle through digital media.

CONCLUSION

Capitalist industries function by controlling the means of production, and the recording industry is no different from other industries. Record labels benefit from selling the cheapest medium possible. Major record labels maintain their dominant position in the recording industry by already having capital invested in the overhead costs of a current medium. However, medium does matter because the production and reproduction of specific media requires capital to produce albums and distribute them to retailers. The means of production in the recording industry consists of copyright, media manufacturing equipment, and distribution networks to get albums to market. Ownership of the means of production creates barriers for smaller/independent labels and new entrants into the market from producing music and getting those albums to retail markets. In fact, the major record labels were once identified as "'major' when they owned their own manufacturing plants and directly controlled their distribution outlets in addition to simply producing records" (Chapple and Garofalo 1977, 15) into the 1950s. Since the '50s, major record labels have relied on their ability to manufacture and distribute music as a mechanism that supports their dominance in the music industry. This creates a tension between a label's ambitions to cut costs to beat its competition while maintaining barriers to new entrants in the market for music because the transformation to a cheaper means of production creates room for new competition as cheaper technology lowers entrance barriers.

The recording industry was significantly affected by the transition from physical media to digital media. This transition in turn changed the form of the music commodity. These changes did not happen to a passive industry that is desperately clutching to media of the past, but rather the major record labels propelled the transition from CDs to digital singles. Therefore, the digital transition resembles previous transitions of recorded music as major record labels seek new means of consumption.

Chapter Two

The Expansion of Consumption in the Recording Industry

At the beginning of the digital transition of recorded music, the recording industry sold CDs. While CDs were the visible artifact of recorded music, they were merely the tip of the iceberg of the production of value in the recording industry. Understanding the evolving nature of the recording industry's commodification of music requires grasping the production of value in the recording industry itself. And the production of value has always been related to the deployment of copyrights. Yet the utilization of copyrights has not remained constant. The ways the recording industry changed what it sells underscore how this value is always changing. Because of the digital transformation of the music commodity, the major record labels actually *increased* music sales and their overall profitability. By developing more sites of consumption, the major labels created more avenues of revenue, as can be seen in shipment reports for the Recording Industry Association of America (RIAA).

RECORDING INDUSTRY VALUE

Despite the constant complaints by the recording industry about declining revenues and profits, it remains both unclear to what degree revenues have declined and unsubstantiated that profits are down even if revenue is down. In an article addressing the supposed decline of recording industry revenue, the International Federation of Phonographic Industry's (IFPI) chief executive, John Kennedy, exclaims, "We're all fed up talking about piracy, it's boring talking about piracy, but it is the problem and we can't avoid it" (Pfanner 2010). According to Kennedy, the rise of digital sales does not offset the decline in revenue from plunging CD sales caused by "piracy." One of the reasons that the recording industry can continue to construct a narrative of financial loss while they appear to continue to increase unit sales is that they do not have to divulge their actual revenues and profits. The structure of large corporate conglomerates makes it very difficult to measure the financial condition of the recording industry. While publicly traded corporations are required by law to report their financial records every year, they do not have to break down their financials by segment, industry, or division (Miller et al. 2000). Whether Sony Corporation is making a profit or posting losses has little or no correlation with how well Sony Music Group is performing. When the record labels, the RIAA, and the IFPI make claims about the viability of the recording industry, it is difficult to verify what they are saying.

Instead of providing information on revenue or profit, the RIAA and the IFPI release data about the U.S. and global recording industry each year, but their data are helpful only in understanding industry shipments—that is, the number of albums "shipped" to retail outlets. Shipments have no correlation to the number of albums purchased or the actual price paid for any given unit (even though both the RIAA and IFPI calculate an average album cost each year). Using shipment data allows the recording industry to provide false financial records by calculating the strength of the industry in their own terms. Several problems arise from this reporting method in calculating the effects of digital technology on music sales.

First, the number of shipments has no real relationship to sales. For instance, a record label could ship one hundred units to a retail store and only sell five albums; the remaining ninety-five unsold albums would be shipped back to the distributors and then to the labels. RIAA and IFPI treat unsold shipments the same as sales of these albums. Before 1992, *Billboard* calculated its charts on shipments and surveys of record store clerks. Record labels would exploit this data collecting system by shipping more albums than would sell at the retail outlets. This would then move an artist up the *Billboard* charts.

Second, corporations have developed increasingly efficient systems, based on point-of-sale technology, that help to identify where a commodity is being sold. This system, combined with faster manufacturing/printing and delivery, allows retailers to stock fewer albums in their stores. Record labels and their distributors can tell where an album is selling as soon as it sells. The point-of-sale system, based around universal product codes (UPCs) and telecommunications systems, helps to ensure that too many units are not shipped to a place where the units are not being sold (i.e., post-Fordism, "just-in-time" manufacturing); a similar process is at work in the clothing industry (Smith 1997b). After a unit is sold out in a given retail store, a label can print and ship albums to that store in a short time, reducing the waiting period for consumers. This process diminishes the uncertainty surrounding album sales. The lower uncertainty means that record labels ship fewer units. Since the record labels began shipping fewer units as these systems developed around the year 2000, RIAA and IFPI data show a reduction in shipments, but this does not mean that labels sold any fewer units; record labels could have been selling more units because their retail process was more efficient at the same time that they reported fewer shipments.

Third, the increase of "big box" retailers affected what music was shipped. Before music became digitally available online, there was a shift from "mom and pop" music stores to chain music stores and later to supermarkets and big box stores (IFPI 2003). Where music stores carry a wide range of music, supermarkets and box stores (Target, Walmart, Best Buy, etc.) stock only the top-selling albums and music by established artists (Hull, Hutchison, and Strasser 2011; Knopper 2009). As consumers become more accustomed to purchasing music at box stores, they become more limited in their music selection because of the limited stock at these stores. Music also has to compete with other media (e.g., DVDs and video games) for shelf space at these stores. The IFPI actually admits that sales "have been affected by competition from newer forms of entertainment, particularly DVD and video games which saw strong growth in 2002. This has reduced the amount of retail space available to CDs and cut into consumer spending on music" (IFPI 2003, 3). The relationship to shipments here is that labels do not have the retail space to ship albums to; this further reduces the number of albums shipped, even though the convenience to shoppers in these stores may increase the number of albums sold by a handful of artists.

Nielsen SoundScan provides more accurate data on music sales than the RIAA and IFPI because it uses UPC data at the point of sale to measure the number of units sold. SoundScan was developed in response to a need for *Billboard* to measure its weekly charts more accurately in the United States. Rather than relying on shipment data and surveys of retail clerks, SoundScan gives accurate data on what is actually sold. As a result, the SoundScan data demonstrate that actual music sold is always less than shipment reports. For example, in 2000 the IFPI reported CD sales of 942.5 million in the United States, while Nielsen SoundScan reported the sale of 706.3 million CDs in that same year (IFPI 1996; Nielsen 2001). Figure 2.1 is a visual representation of the data that SoundScan provides versus the data that the RIAA supplies and exhibits three key details. First, figure 2.1 demonstrates that while the RIAA and the IFPI claim that the first decline of CD sales in the United States occurred in 2001, SoundScan shows an increase in CD sales in 2001 with the first decrease in 2002. Second, according to the RIAA's shipment data, the industry moved 260 million more units in 1996 than in 2007, but SoundScan shows that the recording industry sold ten million more units in 2007 than 1996 (see figure 2.1). Finally, figure 2.1 demonstrates that there is stabilization between RIAA shipment data and SoundScan sales data; this is likely because of a more efficient point-of-sale system. Since the shipment data are closer to the sales data, the record labels are likely earning higher profits on their sales because they have fewer unsold CDs. Furthermore, Nielsen SoundScan does not speculate on the value of the unit sales, which varies greatly depending on the specific unit. By using shipment data, the major record labels can control the story about the condition of the recording industry in the face of digitization.

Self-reporting by the major record labels through the RIAA and the IFPI is a highly problematic measure of actual sales. While the number of album shipments is in fact in decline, other factors could have led to these declines. The digital transformation has provided the major record labels with the tools to argue that their sales are in decline as a result of digital piracy while it has become increasingly difficult to measure the actual revenue of individual record labels.

INTELLECTUAL PROPERTY, NOT FORMAT

During moments of transition, the recording industry is most concerned with reinforcing intellectual property rights. The recording medium is only incidentally important to record labels because record labels sell the content on the medium. As one major record label executive explains, "We're not in the

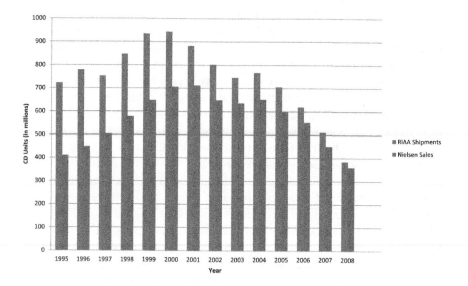

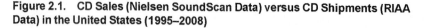

Figure 2.1.   CD Sales (Nielsen SoundScan Data) versus CD Shipments (RIAA Data) in the United States (1995–2008)

CD business, we're in the music business" (Knopper 2012). While consumers buy the physical commodity, the recording industry actually sells intellectual property (in the form of copyright) to consumers. Conceptually, the way that the recording industry functions is not by selling particular media formats (i.e., records, tapes, CDs) to music consumers, but rather by selling the content on the particular medium; as Hull, Hutchison, and Strasser explain, "the recording industry runs on its copyrights" (2011, 51).[1] Because of copyright protections, only a copyright owner can authorize the production and reproduction of a recording in a given format. Record contracts generally stipulate that the record label is the primary copyright owner of recorded works (Hull, Hutchison, and Strasser 2011; Krasilovsky et al. 2007). Through the ownership and control of copyrighted music, record labels generate profit; they control the production and reproduction of the musical content, and by owning copyright, labels can sell content on emerging formats. Copyright law also defines specific things that consumers cannot do with the music that they purchase; most importantly, a music consumer cannot reproduce and sell a copyrighted album without authorization. Therefore, the particular media on which the record label sells music is ancillary to the content contained on the media.

Record labels are not concerned with the actual production or reproduction of physical media, but rather are concerned with the ownership of the copyright of the recorded content. Ownership of the means of production in the recording industry consists of owning copyrighted material more than it has to do with the ownership of studios, manufacturing plants, and distribution systems. Briefly, recording contracts consist of musicians trading their copyrights in exchange for cash advances to produce an album (Hull, Hutchison, and Strasser 2011, 198; Krasilovsky et al. 2007, 14). Copyrights give record labels the ability to create surplus value from their recording artists by transforming artisanal musician labor into wage labor.[2] Selling physical media is a means to an end rather than an end in itself for the recording industry because, conceptually, record labels are licensing the reproduction of copyrighted material rather than reproducing recorded material. Record labels earn profit from the copyright, not from the physical sale; CD manufacturers earn profit from the production of the CDs themselves. During the digital transformation, major record labels have made it clear to the public that they are concerned with copyrights by adapting and enforcing copyright law in the face of "piracy."

Record labels do not generate revenue only from current music; they also benefit from the copyrighted music that they have owned over time. Catalog sales are those "sales of records that have been in the marketplace for over 18 months" (Hull, Hutchison, and Strasser 2011, 248). While albums from deep in a label's catalog may seem to be of little value, a record label's catalog can be important in both rereleasing old albums and licensing for three reasons.

First, some artists remain popular for decades or experience a spike in interest around certain events. For example, the death of Michael Jackson resulted in Jackson being the top-selling artist in 2009, selling over eight million albums in the United States, nearly doubling Taylor Swift's second-place album (Nielsen 2011). This also contributed to a 3 percent boost for Sony's[3] market share in 2009 (Nielsen 2011). Second, since the music is already recorded, there is negligible cost to the label to rerelease an album if an artist's popularity surges. In many respects, a label's catalog has already been tested in the market; reproducing and marketing albums that were previously successful is one way that a record label can ensure a quick injection of cash. Third, this music can easily be licensed to other branches of the culture industry. Music becomes one source of content among many when it is "repurposed" to sell through other media (Caldwell 2004). Licensing music from a label's catalog for use on a TV show, movie, or video game is pure profit for a record label because there is no cost for using the rights to a previously recorded album. This licensing often repurposes content when executives aim to "calculate, amass, repackage, and transport the entertainment product across the borders of both new technologies and media forms" (Caldwell 2004, 50). Catalog album sales come in the form of not only reissued but also remastered albums and greatest hits collections that may include previously unreleased versions of songs. A record label's catalog is a constant source of content that continues to generate profits long after any given song's initial publication. While catalog revenue has always been important for the major record labels, especially during moments of transition, a record label's catalog was especially important during the digital transformation. This is because record labels have developed ways to use new technologies to utilize the copyrights that they own (e.g., by having music performed in video games).

Copyright owners are using new media to cultivate further the ownership of copyrights by licensing the rights to their music across media. Current copyright law creates two types of copyrights that have two corresponding types of catalogs in the recording industry. First, there is the copyright for the author of the composition. This is the stronger of the two copyrights because whenever someone wants to record or perform a composition, the owner of the copyright for the musical composition must be compensated through a license; the royalty rate can be negotiated, or the entity that wishes to record or perform the song can pay the compulsory license rate. After a song is recorded for the first time, "any record company or artist could make a recording of the same song by using the compulsory license," thereby bypassing permission from the original composer (Hull 2004, 58). Songwriters work through publishers to get their songs to performers and audiences. Publishers do not record music (aside from demos); rather, they promote the music that songwriters write to artists, producers, artists and repertoire (A&

R) staff, and other actors in the culture industry (Hull, Hutchison, and Strasser 2011, 119). After the music is recorded, publishers distribute certain royalties to songwriters, having first taken a cut of the revenue as stipulated by the songwriter's contract. Music publishers are not labels, even though the major record labels own the largest music publishers.

Second, there is the copyright of the sound recording. Musicians on a recording who own this copyright receive royalties on the sale (i.e., physical reproduction) or (some) public performances[4] of the actual recording (Hull, Hutchison, and Strasser 2011, 96). Sound recording rights "apply only to the actual sounds captured on the original recording. Thus one could make another recording that imitated or attempted to sound just like the original recording as long as the new one was made by "hiring" different labor to record it (Hull 2004, 50). Record labels are usually the primary beneficiaries of the royalties of sound recordings and disperse any royalties owed to musicians on the recording. When another group of musicians records a song, only the songwriters receive royalties for the recording of the song—the original group of musicians that recorded the song get no royalties from the rerecording (unless they were involved in the writing process) even if they popularized the song; this what is known as a soundalike: "sound recordings that imitate or simulate the original copyright recording" (Krasilovsky et al. 2007, 64). Catalogs give record labels the ability to generate revenue long after capital was invested in recording, producing, and distributing a piece of recorded music. I am highlighting the distinction between types of song catalogs because the tension between composition rights and sound recording rights is fundamental to the way that the recording industry transitioned to digital media.

Throughout the digital transition, the major record labels became more determined to exploit the copyrights that they own through the selling of performance rights; in the United States, performance rights revenues have gone from $9.19 million in 2004 to $70.2 million in 2009 (IFPI 2005; IFPI 2009). A performance right is the "right to perform a work publicly" (Hull, Hutchison, and Strasser 2011, 335). In the relationship between recording and composition, record labels extract profits from the recording of a song in two ways. First, record labels extract profits through their royalties on the sale of or performance of recordings on the label. Second, record labels acquire profits through their corresponding publishers for the authorship of the song. While the utilization of performance rights is a logical business move, the rhetoric still fits the claims of an industry in disarray as a result of "piracy." For instance, a *Rolling Stone* article contends that performance rights "are an increasingly important revenue source in the struggling music business" (Greene 2012). By asserting that the recording industry is exploiting copyrights to make up for lost revenue, increased performance rights revenues appear as a reaction to circumstances outside of the record labels' control. However, the utilization of performance rights was a conscious strategy by the major record labels that preceded any discussion of digital piracy.

Part of this utilization of the catalog resulted in the ubiquitous nature of popular music. The catalog is "a source of income that has become much more significant as the multimedia environment of the late twentieth century offers more and more opportunities for copyright owners" (Hesmondhalgh 2006, 251). Whenever television, advertisements, movies, arenas, or shopping malls play copyrighted music, the copyright owners of both the sound recording and the composition earn royalties from the performance of the music (i.e., performance rights). For instance, television and radio advertising firms used to employ large numbers of jingle writers, but now they staff music directors who buy licenses to use popular music recordings in their advertisements. The amount that the copyright owner earns varies depending on the medium, the popularity of the song/artist, the duration of the clip, and whether it is featured or background music. The IFPI claims:

> Performance rights revenues are one of the fastest-growing and increasingly important revenue streams for the record industry. From bars and gyms, to nightclubs and shops, businesses use recorded music to attract and retain customers and improve productivity. The use of recorded music significantly enhances the value of these businesses and boosts their profits. Some businesses, like commercial radio or internet streaming services, simply could not operate without using sound recordings. (IFPI 2008, 16)

According to the IFPI, businesses that play music are profiting from its performance by making their consumers' experience more enjoyable and ought to compensate copyright owners. Through the deployment of these rights, major record labels have developed new ways to expand consumption using digital technology.

Licenses for copyrighted songs and sound recordings are an important source of revenue for major record labels and publishers. Anytime an entity would like to reproduce copyrighted material, a license must be issued for its deployment by the copyright holder. *Synchronization license* is the term assigned to licenses that allow music to be reproduced in videos (Hull 2004, 57). As labels have sought to exploit copyrights in new media, they have increasingly relied on synchronization rights; from video games to car commercials, every time that a different form of media would like to use copyrighted material, they have to pay for the "sync." These licenses add up to a large source of revenue for the recording industry, but the reporting is generally nonexistent or partial at best. For instance, *Billboard* reported that in 2002 "retail sales of music-generated licensed properties topped $1.5 billion" (Traiman 2003), for everything from music licenses for recordings and songs to licenses of musicians' likenesses. However, there is no accounting for this type of revenue in IFPI's *Recording Industry in Numbers* or the RIAA's database. In 2009, the RIAA began accounting for synchronization revenues and reported over $200 million in this type of licensing revenue ("Year-End Industry Shipment and Revenue Statistics" 2013), but there is no attempt to account for synchronization revenue prior to 2009.

It can be costly to license a popular song by a popular artist for use in a TV show, movie, or video game. *Billboard* reports that Rockstar Games, maker of the popular *Grand Theft Auto* game franchise, "is paying as much as $5,000 per composition and another $5,000 per master recording per track. If that deal applied to all songs, Rockstar's soundtrack budget may exceed $2 million" for *Grand Theft Auto IV* (Bruno and Butler 2008). With an increasing number of video game producers licensing music in the games, the revenue from synchronization is staggering, but again, the RIAA and IFPI have only recently began releasing reports on synchronization revenues. This includes games that have soundtracks, such as the *Grand Theft Auto* franchise, and games that make music the central theme, such as *Guitar Hero*, *Rock Band*, and *Dance Dance Revolution*. Def Jam Records has been particularly aggressive at courting video game makers to license music from its catalog (Ault 2002); these include music games such as *Def Jam Rapstar* and sports games such as *Def Jam Wrestling*. However, major record labels, on the whole, are not happy with the rate at which they are licensing music (Bruno 2008a). At the same time, there have been bidding wars between Activision (*Guitar Hero*) and MTV Games (*Rock Band*) over access to catalogs for popular recording artists such as the Beatles, Aerosmith, AC/DC, and Metallica (Bruno 2008b). Needless to say, the recording industry recognizes that there is a lot of revenue to be earned from utilizing label catalogs in video games.

Sometimes the songwriter and publisher will license the recording of a song and circumvent an expensive recording artist; such circumvention is evident in the video game *Guitar Hero* when unrecognizable bands are listed next to popular songs (additionally, the song sounds different). This happens when the songwriter is paid a synchronization license to have the song on the game, but the original performer was not paid for the sound recording of the popularized recording of the song. As long as a video game company pays the songwriter for the recording, then the video game company does not violate copyright when it records a soundalike with its own performers. For instance, if *Guitar Hero* wanted to put Michael Jackson's "Thriller" on the newest version of the game, Jackson's estate would likely charge a large fee. However, Rod Temperton is the songwriter, so Temperton (and his publisher) can authorize another group to "cover" "Thriller" for the purpose of performing it on the video game. *Guitar Hero* would have to pay Temperton either way, but the use of a cover version would be noticeably cheaper. This technique is not without controversy as some bands have sued *Guitar Hero* claiming that the songs are too similar to the original works, as happened with the band the Romantics (Butler 2007; McCollum 2007). While recording artists fight this apparent infringement, record labels, and their respective publishers, support the use of these songs.

In a further development, Sony/ATV (the world's largest music publisher and partner of Sony Music Entertainment) began hiring unsigned artists to record cover versions of popular songs for its YouTube channel entitled *We Are the Hits*. The idea was simple: Sony/ATV could rerecord popular songs that it already owns the rights to and create music videos for the songs. By skipping recording artists, Sony/ATV created a strategy to generate additional revenue; "because the covers are not original recordings, the only rights to be paid are those of the songwriters" (Adegoke 2014, 15). This creates a stream of revenue for music publishers that not only bypasses recording artists but also allows publishers to enter into relationships with performers. Since Sony/ATV dictates the amount that they pay these singers, they increase the revenue that the publisher and songwriters would otherwise receive from recorded music. The "nearly 130 million views per month" (Adegoke 2014, 15) that *We Are the Hits* averages on YouTube expands the consumption of recorded music beyond the realm of possibilities available in a physically mediated music system.

Labels and publishers have large profit margins on licensing because there is no cost for the physical media, distribution, or marketing of the music. By owning the rights to music, record labels can continue to extract profit every time that there is limited interest in using a song in the future. As long as labels do not have to manufacture and distribute physical media as they do with mechanical licenses, there is no cost to the label. Furthermore, since copyright detaches music content from media format, catalogs allow record labels to reprint in new media. Ownership of copyrights gives record labels the ability to constantly adapt to new media and use music in other cultural industries while reaping the benefits of recorded music that has already had its cost recouped.

INDUSTRY-DRIVEN DISINTERMEDIATION AND COMMODIFICATION

With starkly different numbers coming from the IFPI/RIAA and Nielsen SoundScan (see figure 1.2 and figure 2.1), it is important to call into question the connection between piracy and decreased CD sales. If fewer CDs are being sold, what has caused the decline in CD sales? Arguably, music consumers began to rely more on digital music formats in the 2000s, but is this the only factor that caused a decline in CD sales? One contributing factor, often overlooked by the recording industry, involves the album replacement cycle, but even when the recording industry acknowledges this deficit, industry representatives continue to claim that the problem is "piracy." Is piracy the only cause of the decline in physical sales? As the Internet has made online digital distribution feasible on a mass scale, the major record labels have accelerated the decline of physical media formats, especially the CD, and created a number of new commodities for music fans to consume.

Major record labels compelled the change from CDs to digital music files. While the number of CDs sold has drastically declined, the sales data demonstrate a historic rise of digital music sales since 2003. In order to account for digital sales, Nielsen added the category of overall music sales to its data service in 2004. This shift was necessary because measuring only the sale of physical media and albums was no longer accurate with the rise of digital track sales. Figure 1.4 in chapter 1 shows the trend of overall music sales (1995–2010). The move to overall music sales demonstrates that online music retail has been a major success. According to Nielsen SoundScan, overall music sales in the United States were at 1.369 billion in the year 2007 (the year that corresponds to the 2008 *Recording Industry in Numbers* report); this means that actual sales of recorded music were at record-breaking levels in 2007. Overall music sales in the United States have been over one billion units per year since 2005 (see figure 1.4). Aside from a slight decline in 2010,[5] overall music sales have increased every year since SoundScan began measuring the category in 2004.

However, the IFPI and the RIAA maintain that music sales would be even higher if it were not for music pirates. "Music consumption," the IFPI claims, "is at an all-time high and continuing to grow at a healthy rate. Music remains a top leisure activity and yet consumers are spending less on music as the use of unlicensed free channels continues to outstrip legal music acquisition" (2008, 12). The IFPI provides no data to demonstrate that "piracy" outstrips actual music sales in its reports. According to a report cited by a marketing research firm on the RIAA's website, only 37 percent of music was paid for between 2004 and 2009 (Graham 2009). If the research data on piracy are accurate, then the recording industry claims that it should have sold roughly 3.7 billion units in the United States in 2007. This estimation is based on surveyed listening practices (Graham 2009) that assume that any and all acts of listening are also discrete acts of consumption. In previous media, neither a tape cassette exchanged between two friends, nor a person listening to music on the radio would be considered consumption in these terms. However, the RIAA and the IFPI want the public to believe that they must pay to listen to all music to be legal—and as I will show in subsequent chapters, they are changing the law and technology to enforce this perspective. With digital rights management, the recording industry is trying to commodify all listening practices.

During the digital transformation of music, the recording industry has broadly expanded the means to consume music. As Paul Smith argues (expanding on Michel Aglietta), the explosion in growth of leisure commodities constitutes "one component in capital's response to the crises of the postwar period" (1997a, 48). Finding new ways for music fans to consume music is one way that the recording industry has responded to general crises of capitalism. "The recording industry has reinvented itself," according to the IFPI, "moving from a one product business to a multi-product, multi-channel digital business" (2007, 3). While making digital music available online through Internet retailers, such as iTunes, is an obvious new way both to sustain and to increase music consumption, digital albums and singles are the tip of the iceberg of the music commodities available today. There has been an increasing attempt on the part of record labels to generate revenue through performance rights, and by licensing music for video games or playing music at sports arenas. The IFPI acknowledges that although the recording industry did not report music licensing as part of its revenues in 2003, "this may need to change in the future as mobile phones and video games use recorded music as an integral part of their offering" (IFPI 2003, 3). There is also a move on the part of the recording industry to make music a "lifestyle"; with "'lifestyle' music, consumers are becoming exposed to music via film, television, games and mobile devices" (IFPI 2004, 3). These new sources of revenue for the recording industry require the exploitation of performance rights.

Aside from digital music stores, some of the new formats are as follows:[6]

- **Master ringtone**—A master ringtone is a cell phone ringtone that plays an actual song. It allows consumers to play a single when someone calls. Some people have multiple master ringtones for different callers. *Billboard* has charts for the top-selling ringtones. In 2004, for instance, $2.5 billion of revenue was generated by the recording industry from ringtone sales (Flynn 2004). However, the popularity of ringtones has decreased over the past several years.
- **Ringback tone**—A ringback tone is the music that is played for a caller that functions similar to an on-hold song.
- **Internet radio**—Internet radio provides listeners with a continuous stream of music; this music is not downloaded. While terrestrial radio does not have to pay mechanical licensing fees for the performance of copyrighted music, Internet radio stations (such as Pandora) are required to pay performance rights for every "spin" of a song.
- **Subscription services**—Users pay a fee to a subscription service in return for access to the service's music library (e.g., Napster, Rhapsody, Spotify, Beats). As long as subscribers continue to pay the fee, they can listen to music, but as soon as they quit paying the fee, they lose access to any music that they downloaded. Figure 2.2 shows the considerable rise from $6.9 million in 2004 to $461.6 million in 2010 of subscription services in the United States. According to data from the RIAA, the recording indus-

try generated over one billion dollars from subscription services in 2012 ("Year-End Industry Shipment and Revenue Statistics" 2013), and this sector continues to increase exponentially. There is very little overhead for this revenue, and once the consumer begins paying for a monthly subscription, they are likely to continue the service indefinitely.

- **Ad-supported websites**—On ad-supported websites, users get to choose the songs that they want to listen to, similar to subscription services, but the content is paid for through advertisements. These sites differ from Internet radio because users have control over what they listen to, so they do not follow the same licensing rules. Examples include Spotify, Last.fm, Allmusic, and MySpace.
- **Video-sharing websites**—Video-sharing websites allow users to upload videos onto their servers. When users upload videos, the videos often include copyrighted music. The host website must pay copyright owners for the content (unless the copyright holder orders that the video be removed from the site), and this usually paid for through advertising. Sometimes record labels upload their recording artists' videos onto these sites and use them as both a source of revenue and a source of advertising. Examples include YouTube and Veveo.
- **Video games**—Video games incorporate music in two ways. First, video games like *Need for Speed* or *Grand Theft Auto* incorporate music into the game. Second, music video games, such as *Guitar Hero* or *Dance Dance Revolution*, are based around gamers performing music. Video game companies must pay copyright owners licensing fees to incorporate music into games.
- **Apps**—Applications (known as "apps") are closed-web interfaces (Zittrain 2008) that allow users to perform specific functions on the Internet. Record labels are developing new ways to use apps to deliver music through the Internet. Apps can be used to deliver albums with specialized content that is available only through the app.

Additionally, music consumers can stream content on the Internet in a number of ways that are supported by advertisements. This new revenue is rarely reported in the data that contributes to figure 1.1 and is part of the lack of transparency on the part of the recording industry discussed in the introduction. The RIAA and the IFPI conceal the added revenue that they gain from these streams at the same time that they complain about the impact of "piracy." While the primary commodity of the recording industry remains recorded content, the recording industry has expanded the ways that people consume music during the digital transformation of music.

In the past, new media have given rise to new industries while competition across and between industries has resulted in the decline of older industries. However, today the recording industry insists that it must see increases

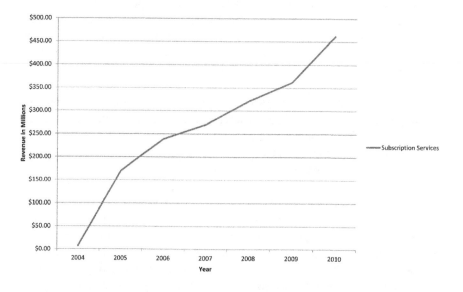

Figure 2.2.   Revenue from Subscription Services according to the RIAA ("Year-End Industry Shipment and Revenue Statistics" 2013)

in profit even as its position in content production has changed; in other words, major record labels feel that they are "too big to fail." Interestingly, in 2008 and 2009, when the U.S. economy was in the throes of a massive recession, the recording industry increased its overall music sales by more than 176 million units, as demonstrated by figure 1.4. Additionally, the IFPI reports that U.S. performance rights revenue for the recording industry more than doubled from $23.5 million in 2007 to $54.8 million in 2008 (IFPI 2008). While CD sales did decline over this period, there is no reason to think that CD sales would not decline. Because of disintermediation, digital music and online distribution allow major record labels to circumvent most of the cost of manufacturing, storing, and shipping CDs, along with cutting the cost of intermediary retailers. By reducing the costs of physical production and distribution, record labels are able to reduce their prices and leave consumers with more money to purchase more music. Music consumers purchase singles instead of albums in part because record labels emphasize the sale of digital singles and encourage a corresponding move away from CDs; this is no different from the recording industry's move away from tape cassettes in the 1980s. While the end of the album replacement cycle and the availability of free files on P2P networks may have contributed to the decline of the CD, the major record labels propelled the format's phase out.

HARDWARE VS. SOFTWARE

We must also consider the relationship between software and hardware during the digital transformation of music because the companies that create content often have divergent interests to those that create the devices that play that content. Mediation is an important topic because recorded music has to be mediated through technology; in effect, technology produced by consumer electronics manufacturers structures the relationship between listener and music. The recording industry is dependent on the consumer electronics industry to dispense music to its consumers, while the consumer electronics industry is dependent on the recording industry to produce content that can be played on electronic devices. Though this relationship is often symbiotic, at times the consumer electronics industry tries to promote a new technology that the recording industry is concerned will interfere with its business model.

Because of the industrialization of music, technology and music are difficult to separate. The recording of music is so connected with popular music today that it is hard to conceive of music without thinking of the recording industry. Simon Frith contends that the "industrialization of music cannot be understood as something which happens *to* music, since it describes a process in which music itself is made" (2006, 231). Contemporary popular music is the product of a process that cannot be separated from its position as a recorded object; "from a historical perspective, rock and roll was not a revolutionary form or moment, but an evolutionary one, the climax of (or possibly footnote to) a story that began with Edison's phonograph" (Frith 2006, 232). In this way, popular music is connected to the technology through which it is mediated. As long as we speak of the "recording industry," music is being understood specifically as a recorded commodity.

Frith notes that the development of the recording industry is a product of the relationship between "hardware" (i.e., the music player) and "software" (i.e., the content played). When gramophone manufacturers began mass-producing gramophones, they needed something to sell to play on the gramophone (Frith 2006). The companies that produced the hardware were the first producers of software and became today's record labels (e.g., Columbia Records). Whereas the classical music field developed new technologies to produce high-fidelity recordings, pop music "developed recording as an art form" (Frith 2006, 238). This gave rise to the development of producers and technicians as artists in their own right.[7] Furthermore, as new recording technologies develop, Andrew Goodwin contends, we must also look at the ways that "the actual conditions of production" change (1992, 97). Specific political-economic forces bring about changes in technology, but they are neither inevitable nor optimal just because they are new; rather, it is important to understand the conditions that drive technological innovation.

New media threaten to disturb the balance between consumer electronics manufacturers and record labels because consumer electronics manufacturers seek new technologies without regard for the effect (positive or negative) that they could have on other industries. With each new device, consumer electronics manufacturers:

> feel entitled to capture all of the new value that their innovation makes possible. The creators of music (and their investors) make the point that without the music the new service or product would be valueless. The two sides eventually realize they need one another, and a balance is found. (Krasilovsky et al. 2007, 414)

Consumer electronics manufacturers make their revenue from inventing new ways for their consumers to listen to music, while the record labels generate revenue by producing new music for their consumers to listen to; this is the fundamental relation between these separate industries. In order for consumer electronics manufacturers to increase their rate of profit, they must develop, patent, and produce a new way to listen to music that their consumers can purchase from them before their competitors develop an equivalent or better technology. After producing the new music listening device, consumer electronics manufacturers generally have to entice record labels to produce music in the new format. Without the acquiescence of record labels in producing music in the new media format, the new devices are useless because mechanical copyrights stipulate that only the owner of those rights can permit the reproduction of copyrighted material. If a record label refuses to have its music manufactured in a certain format, consumers of that device will be unable to listen to music by recording artists on that specific label. While developing the means to produce a new media format for record labels can be costly, investing capital in new media can also result in increased catalog sales in the new format through the album replacement cycle. However, sometimes the interests of the two industries are divergent and the tensions boil over into lawsuits, as they have during the digital transformation of the music commodity.

In the early years of recorded music (1890s–1930s), there was little conflict over the reproduction of music in new media formats between record labels and manufacturers because the record labels were also the consumer electronics manufacturers.[8] Early gramophone and phonograph manufacturers were in need of content for consumers to purchase and play on their machines; since there was no recording industry at that time, these manufacturers began recording all kinds of content, music being only one of them (Frith 2006; Kittler 1999). The problem at that time was that the different record manufacturers did not create machines that were compatible with each other; for example, in the early days of the recording industry, albums recorded by Berliner Gramophone could be played only on a Berliner Gramophone. As time passed, the record players became more compatible across companies, but until the 1950s, they still varied to a degree that affected sound quality. In 1952, the RIAA was created to administer an equalization curve to ensure higher sound quality across players (Hoglund 2011). After the formation of the RIAA, record labels began to focus more on content than hardware—even though format disputes sometimes arose, as with, for example, the commercial release of stereo in 1957 by EMI (Chapple and Garofalo 1977, 53).

While consumer electronics manufacturers and record labels often fall under the same corporate conglomerate today, these independent branches of the conglomerates rarely work together. The main goal of these conglomerates is to satisfy shareholders and increase the value of the conglomerate's stocks (Chapple and Garofalo 1977; Harvey 2005). Often, larger electronics conglomerates purchased record labels in the hope of developing a new media format under the assumption that they could use the label to license music to play on the new media player (Hesmondhalgh 2007, 166). However, large conglomerations can also develop "competition among product line divisions" (Mosco 2009, 160). For instance, Sony Music Group is an entirely separate entity from Sony Electronics. If Sony Electronics decided that it would profit from producing a device to play Internet radio in cars, but Sony Music Group was opposed to the idea of portable Internet radio, then Sony Electronics would still produce the radio as long as being the first on the market boosted Sony's stock values. While it may seem at first glance that two companies in one conglomerate would want to work together, they actually work independently for the benefit of the conglomerate as a whole.

Digital music has exacerbated tension between the recording industry and the consumer electronics industry because digital technology allows for the home reproduction of music without any loss of fidelity as a result of copying—something that was not a concern for analog formats. Since digital files can easily be reproduced, record labels, at least initially, lost their ability to restrict the adaptation of their music into new formats. This tension first arose from the creation of the digital audio tape (DAT) recorder. DAT recorders gave users the ability to make copies of copies without any loss in fidelity. The fear of the recording industry was that widespread commercial availability of these devices would lead to widespread unauthorized reproduction of copyrighted music (i.e., "piracy"). Congress passed the Audio Home Recording Act (AHRA) of 1992 to deal with this issue; the AHRA in turn mandated the inclusion of digital rights management (DRM) that created loss of sound fidelity for copies of copies. The AHRA also stipulated that consumer electronics manufacturers pay royalties to the RIAA for blank digital recording media designed for recording music and for digital recorders like CD burners (Hall 2002). While the AHRA established processes and mechanisms that appeased both industries with regard to DAT recorders, the AHRA did not address the problems that would arise from future digital innovations for recorded content.

When the first MP3 player became commercially available in the late 1990s, the recording industry moved to prevent production of the devices. Three years after the first release of MP3 files, Diamond Multimedia released the Rio PMP300 in 1998, which allowed users to listen to portable digital music similar to the Sony Walkman or Sony CD Walkman. The RIAA immediately filed a lawsuit seeking an injunction from the courts to stop the production of the MP3 player (Bate 1998). At the time, the RIAA argued that every MP3 file was an illegal copy because of the lack of any "legitimate" MP3 retailers; by that argument, the Rio PMP300 was capable only of playing illegal files. However, the courts disagreed and continued to allow manufacturers to produce MP3 players (Starrett 1999). This moment is important because it signals a shift in power between the consumer electronics industry and the recording industry as consumer electronics manufacturers were no longer dependent on agreements with the recording industry to create content to play on their devices. Whether or not record labels agreed to have their music reproduced in digital files was irrelevant to whether or not music listeners would have music to play on digital devices. Consumer electronics manufacturers circumvented, at least temporarily, the record labels' control of content in a digitally mediated musical environment.

CONCLUSION

Over the past two decades, the recording industry has been going through a profound transition from analog to digital music. While the recording industry, through the RIAA and the IFPI, repeatedly argued that "piracy" was causing the industry to hemorrhage, this chapter demonstrates that there is reason to pause and further scrutinize the recording industry's claims of financial loss. The recording industry intentionally shifted its focus from CDs to digital files just as it shifted its focus from tape cassettes to CDs in the 1980s. Furthermore, the recording industry's condition needs closer examination because by all measurements it vastly increased the overall music sold between 2000 and 2010 and at the same time increased its utilization of performance rights and synchronization licenses across other cultural industries. Increased efficiency comes to record labels from reduced costs in production, manufacturing, and distribution. Record labels can produce fewer songs, sell music online, and double down on their deployment of songs in their deep catalog; all of these tactics result in record labels investing less capital in the music commodity while at the same time selling more units than ever before.

As the music commodity changed from analog to digital, the nature of the recorded product fundamentally changed as well. Albums became less popular as the popularity of the digital single has skyrocketed. The move from albums to singles, led by the recording industry, made recording more efficient because there is no longer a need to produce filler for the longer albums. Record labels now have the option to produce only the music that they, and their recording artists, feel will sell reasonably well. While music fans worry that this is fundamentally changing music as an art form, it is important not to reify past media formats of the music commodity. When the nature of the music commodity changes, what actually happens is a change in the means of production advanced by capitalist interests.

Even more important than the details of the changes in the recording industry is the way that the recording industry managed to maintain its importance in the larger music industry even as the shift from industrial production to digital production threatened to make record labels obsolete. The major record labels convinced the U.S. government and the American public that their dominance in the music industry is necessary for the continued existence of music. The next part of the book scrutinizes how the RIAA used the U.S. government to create laws that allow the major record labels to maintain their dominant position within the larger music industry.

Part II

The State in Music

Chapter Three

Copyright: A Critical Exploration

Few people understand copyright, but everyone seems to think that they comprehend it. The problem is that copyright is not founded on some broad ethical paradigm, but rather it is the result of a series of negotiations over several hundred years among stakeholders. When the recording industry makes appeals to the public using the piracy panic narrative, they exploit the fact that the public thinks they understand copyright as a basic property right. As a result, most people fall back to a simple position that stealing is wrong. But what if violating copyright isn't stealing? What if downloading music isn't even violating copyright? Where does that leave the piracy panic narrative? Who does copyright benefit? Who does it exploit? These are questions worth exploring, but the piracy panic narrative generally clouds a frank discussion about the ends of copyright.

In the relationship between the state and capitalism, copyright is a mechanism, defined through law by the state, that establishes a very specific kind of production in the recording industry (and music industry, more generally) by creating a legal apparatus that allows the industry to produce revenue. Therefore, it is highly important to look at the conditions under which copyright policy changed during the digital transformation of the music commodity. At moments of media transition, the recording industry has successfully convinced the state to change the law. As a result, I argue that these large corporations play a greater role in getting the government to change the law to meet their needs than what liberal democratic theory contends about the way a democracy should function.

The state is involved in the recording industry first by assigning monopoly rights to recorded content through copyrights. By establishing the ownership of music through copyrighting, constructed as a form of intellectual property, the state creates the means through which value can be produced from the production of music. From sheet music to digital files, copyright is what enables the capitalist production of music. Copyright ownership is transferable, and it is through the sale of musical copyrights by the original owner (i.e., musicians) to record labels that the recording industry creates profit. It is the state's legislation that structures the relationship between record labels and musicians—capital and labor.

The second area of state involvement in the recording industry is in the regulation of the distribution and circulation of commodities. Distribution is characterized by the regulation, or lack thereof, of the systems that get commodities from production to consumers. This can happen at the local level where city councils and county boards can prohibit the building of a large box store (i.e., Walmart, Best Buy, or Target) that ultimately sells music as a loss leader.[1] Or it can happen at the federal level with the Federal Communications Commission's (FCC) regulation of Internet service providers (ISPs). As the medium of music has changed, the distribution of music has also changed. Since distribution directly affects the nature of the commodity,[2] the policies that regulate distribution affect the form of recorded content.

COPYRIGHT

To the extent that the state creates and protects property rights, the state is always already in support of capitalism. This is ideologically established in the Declaration of Independence, which appropriates John Locke's view that government should protect "life, liberty, and property,"[3] but protecting property is also one of the fundamental roles of liberal democracies. Furthermore, protection of intellectual rights is one of the few powers that the U.S. Constitution delegates to Congress. The Constitution states that one of the roles of Congress is "to promote the Progress of Science and useful Arts, by securing for limited Times to Authors and Inventors the exclusive Right to their respective Writings and Discoveries" (Article 1, Section 8, Clause 8). Copyright law, along with other intellectual property rights, is founded on the idea that an author's work must be protected for "limited Times" in order "to promote progress." This is one of the few areas regarding which the Constitution gives Congress the power to legislate, and Congress has continually updated copyright law since the acceptance of the Constitution. However, by understanding this clause as a "property" right, and thereby attaching money to this right, legislators have interpreted that the Constitution solidifies a genial relationship between capitalism and the state for the culture industry.

By creating copyright as a *property* right, the state codified the selling, buying, and hoarding of copyrighted material. When an author participates in the production of a work of music, there is no necessary connection to the imperatives of commerce, but under a regime of copyright law, the author thinks that they "own" the rights to the song. This is a rhetorical position used by the recording industry to perpetuate the notion of artistic autonomy more than a legal position. In fact, William Patry explains, "copyright has never been regarded as a property right. Instead, copyright has always been a regulatory practice granted by the grace of Congress" (Patry 2009, 110). But Patry continues by examining how the rhetoric that describes copyright as property changes the social relations in the system. The fact that the rhetoric about copyright as property is articulated allows people to conceive of it as such; this in turn changes the social relations of production. When a song is owned, the owner can sell (both the copyright to and the use of) that song to others; however, musicians signed to recording contracts typically agree to sign the copyrights of their work over to their label. In effect, the state enables record labels to own their artists' music.

Capitalism is built on the ownership of property as owners restrict their workers' use of, and ability to profit from, that property. In agrarian forms of production, this means that people who own the land can pay wage laborers to work on their land and reap the fruits of their workers' labor. In industrial production, the owner of the factory, equipment, and raw material can pay laborers to work in the factory to produce commodities, the ownership of which reverts to the capitalist. The recording industry functions first by convincing musicians to sell their copyrights in return for an advance. After a label owns an artist's copyrights, labels stipulate under which circumstances those musicians can record and perform their music. In this way, copyright establishes a wage relation between musicians and record labels. While the fundamental idea behind copyright has remained the same, the details of this property relation have evolved over time.

Copyright creates an enclosure that is similar to the property enclosures enacted in Britain from the fifteenth to early nineteenth centuries. As Marx illustrates in *Capital, Volume I* (1992), the enclosure of communal property in Britain allowed the landowning class to extract labor from non-landowning laborers because peasant farmers were forced to work for the landowners in order to subsist. In return for their labor power, peasant farmers received a wage with which they could pay rent and buy commodities. By connecting a peasant's labor power to a wage, capital created surplus value through purchasing the peasant's labor power for less than the value produced by the worker. Property ownership—created by land enclosure acts—thus commodified peasant labor power by separating the worker from the means of production.

Similarly, in the decision to sell their rights in exchange for a record label's advance, copyright separates musicians from their means of production—their creative songs are the object of their labor. Copyright legislation enabled musicians, their estates, or representatives (i.e., publishing companies or record labels) to own previously noncommodified pieces of music.

Copyrights "enclose" a public good[4] by delimiting the boundaries of a song and identifying an owner who can decide legitimate uses of their song. At first, copyrights seemed questionable because most songs were folk art and their creation was not attributed to a specific individual but rather were held in common by the public (i.e., in the public domain). After copyright "enclosure," musicians were able to lay claim to songs that either were not original or were derived from another song. Different media industries also exhibit this process; for instance, Disney is known for producing movies based on stories (i.e., Grimm's fairy tales) that were part of the public domain. After Disney produces the movie, it receives the copyright for its version of the story/movie. Then, by getting the U.S. Congress to extend the length of copyright terms, Disney has been able to hold its copyrights for the movies and stories indefinitely (Lessig 2004). Copyright "enclosures" provide a way to demarcate the public domain and generate profit from the new property form.

When a musician writes a song, the piece possesses no monetary value for which it can be exchanged—music in this sense is not a commodity. Without copyrights, anyone else can perform/reproduce the piece outside of any monetary system. The song has no monetary value until it is considered property. Property, on the other hand, can be exchanged. Law generally defines the taking of someone else's property without permission as stealing. However, property laws do not apply to the ownership of the ideas contained in music because people cannot physically take a song (Green 2012; Lessig 2004; Litman 2003). Rather, copyright intervenes to place a restriction on the reproduction and performance of a copyrighted song, thus allowing its commodification.

Through copyright, authors/writers/composers/artists thus become petit bourgeois producers (similar to small-business owners) of a commodity, but the degree to which they can distribute that music is limited without access to capital because it takes capital to record, produce, and distribute music. Copyright allows authors to transfer the ownership of their music to corporations (in the case of music, record labels) in order to circulate their music to a wider consumer base. Copyrights are an important part of the means of production in the recording industry because they are both the fundamental source of property in music and the object of labor power (Marx 1992). When composers and musicians sell the ownership of their copyrighted music to the recording industry, they are selling their ownership of the means of production.[5]

Since music is a nonrivalrous good (one person's use does not diminish the use of another person), one person's consumption of a song does not preclude another person's consumption of that same song. Whether one person listens to a song on a record player, plays the song over a PA system at a party, or gives it to a friend, the same song can be repeatedly replayed. The same scenario holds true for the performance of a song—when a musician performs a song, other musicians can still perform the song. Copyright interrupts this perpetual performance of music by giving the owner the authority to control how, when, and where their composition can be performed. It also regulates the ways that music can be commercially exchanged. Copyright law creates the property rights that allow for the commodification and commercialization of music.

Copyright legislation creates two types of copyright for music in the recording industry. First, there is the copyright for the writing of the song, which protects composers/authors. Second, there is a copyright for the specific recording of a song, which is the right of the musicians that actually record a song. While both types of copyrights originate from musicians and authors, record labels and publishers own most of these copyrights. The distinction between these different types of copyrights is important because the way that these copyrights are allocated is contingent on the balance of power among copyright stakeholders at the time when Congress updates copyright legislation.

Record labels produce profit by purchasing the copyrights of musicians and composers in exchange for an advance to record, produce, release, and market their music. The signed artists are then compelled to work for the label in order to pay off the advance and record more albums (Hull, Hutchison, and Strasser 2011, 198). It is through copyright that the relations of production in the recording industry are established because copyright law creates intellectual property and regulates the relationship between capital and labor by transforming musicians into laborers through the signing away of their copyrights to record labels. Without copyright legislation, there would be no way to extract profit from the production of music; recorded music itself would have no monetary value. In turn, there would be no need for record labels to reproduce recordings because there would be no way to profit from music.

Because of the commodification that intellectual property creates, copyright legislation is the most important legal tool for the recording industry to maintain. Since copyright enables the relations of production in the recording industry, the state must update the law to allow capital to continue to expand by exploiting labor. As the mechanisms for reproducing music change, copyright owners must actively change the structure of copyright to ensure that they protect their property in new media environments. The digital transformation of the music commodity threatens to disturb the means of production in the music industry; this jeopardizes the major record labels' dominant position in the production and commodification of music. Controlling the reproduction, distribution, and consumption of music through copyright law is the only way for the recording industry to maintain its viability at moments of transformation. During the digital transformation, the recording industry has argued that sharing copyrighted materials is similar to stealing from recording artists (Green 2012); record labels attempt to link the working conditions of their recording artists to the consumption behavior of their fans. However, the copyright property regime enables the exploitation of recording artists' labor by record labels.

COPYRIGHT NEGOTIATIONS

From time to time, Congress is compelled by copyright owners to change the current copyright regime, typically to adapt to new media innovations.[6] However, such periodic revisions are rarely vetted by news media or discussed at town hall meetings. Rather, copyright law is the product of multi-party negotiations among those industries that historically have had a stake in copyright legislation. As Jessica Litman explains, "Congress got into the habit of revising copyright law by encouraging representatives of the industries affected by copyright to hash out among themselves what changes needed to be made and then present Congress with the text of appropriate legislation" (2006, 23). Whereas policy scholars often discuss the involvement of lobbyists in the writing of legislation, Litman points to a much deeper involvement whereby Congress allows some affected industries to negotiate and write the law. Because industry insiders write copyright legislation, this affirms the state's involvement in the economy explicitly on the side of capital. At this moment of transformation from a physical medium to a digital medium, it is important to understand the process through which copyright law changes in order to reveal the impact of copyright modifications during the digital transformation.

Since copyright stakeholders write copyright legislation, multiparty negotiations benefit capital in very specific ways. Legislation written by stakeholders in multiparty negotiations has three features:

1. "No affected party is going to agree to support a bill that leaves it worse off than it is under current law."
2. "There's a premium on characterizing the state of current law to favor one's position, since current law is the baseline against which proposals are negotiated."
3. "The way these things tend to get settled in the real world is by specifying" negotiated privileges. (Litman 2006, 23–24)

These features establish a framework for what the legislation will look like in the end—that is, legislation that creates very particular exceptions that aim to give different stakeholders a footing in copyright. If current copyright stakeholders are unwilling to leave the negotiation table worse off, and if we presume that some interests are antagonistic in any such negotiation, then one can only expect those stakeholders that are excluded from the negotiation table to suffer because of new legislation. Therefore, it is imperative to look at those groups *not* at the negotiation table in order to understand what stakeholders gain as a result of new copyright legislation.

During the last major round of copyright negotiations, around the passage of the Digital Millennium Copyright Act (DMCA), the negotiations included two broad groups of stakeholders: "content owners" and "user interests." As Litman describes, the

> "serious" negotiations—the ones perceived to be necessary to ensure the enactment of legislation—involved the motion picture industry, the music recording industry, the book publishers and the software publishing industry on behalf of the "content owners," and the online and Internet service provider industry, the telephone companies, the television and radio broadcasters, computer and consumer electronics manufacturers, and libraries representing the "user interests." (Litman 2006, 126)

The distinction here between "content owners" and "user interests" is not as significant as their names suggest because these interests are already embedded in the negotiation process. User interests do not include the public at large, even though librarians often try to stake claim to the public's interests. Furthermore, the content owners are represented, at least in the case of the recording industry, by trade associations. Musicians unions and guilds were not included in these negotiations. Furthermore, there is some overlap because, for instance, Sony's interests would be represented by both consumer electronics manufacturers and the music industry. The system through which the copyright law is negotiated is stacked in favor of corporate conglomerates.

Because of a copyright negotiation system that only includes industry stakeholders (identifiable under current copyright law), three key groups are left out of the copyright bargaining table: (1) new industries, (2) the people/public, and (3) musicians/content creators. As new media environments evolve, the difficulty increases for current stakeholders to maintain their position as new media create gaps in the stakeholders' copyright protections. New media result in legal question marks that the current stakeholders would like to have resolved either in the courts or in Congress. Because one set of stakeholders does not want to lose ground to another group, all stakeholders (those currently recognized under copyright law) agree to come to the bargaining table to negotiate a new deal. By leaving these three groups out of copyright negotiations, each group's position in a new copyright regime changes in myriad ways.

First, since new media industries have yet to develop an obvious stake in the law, they are left out of the negotiations. Litman argues that "any successful copyright legislation will confer advantages on many of the interests involved in hammering it out, and that those advantages will probably come at some absent party's expense" (Litman 2006, 62). In this case, the absent party is the newly created industry. The exclusion of new media industries from the negotiations allows current stakeholders to create copyright legislation that divides the rights to the new media without consideration for new industries. Once a new copyright regime is established, those new industries that are not included at the negotiation table have no assurance that they will be included in the future because they are not stakeholders identified in the updated copyright law.

Second, all discussions of stakeholders are incomplete without mentioning the role that the people or the public have in the democratic construction of state policy; however, citizens are not represented at the copyright bargaining table. When copyright legislation is the product of multiparty negotiations, it is "overwhelmingly likely to appropriate value for the benefit of major stakeholders at the expense of the public at large" (Litman 2006, 144). While stakeholders negotiate deals so that they do not lose anything to each other, they absolutely do not want to lose anything to the public. Both Robert McChesney and Litman are concerned by the lack of public involvement in the media policy process. Today, the only time that the public is included in policy discussions is in their role as consumers. As with most aspects of neoliberalism, citizens are supposed to exercise control over copyright policy only as consumers, since in neoliberal ideology a free market ultimately supersedes democratic demands—citizens get their voice in the free market by the commodities that they purchase or abstain from purchasing. The one dimensionality (Marcuse 1991) of this conception of the people in a democracy positions the public as peripheral to politics. Furthermore, Lawrence Lessig argues that legal mechanisms such as "fair use" are critically important for future cultural production (2004; 2006). The public has a number of interests stemming from the Constitution and Supreme Court rulings that are important to uphold during copyright law updates. These public interests are trampled because the public is not included in the copyright negotiation process.

Historically, the public has been accorded certain rights within the realm of copyright that limit the absolute monopoly that copyright owners have over content. The following doctrines are held to be in the public's interest, but these interests are rarely included at the copyright negotiation table. As Lessig (2006) describes it, the public interest has been taken into consideration through (1) the "fair use" doctrine, (2) the "first sale" doctrine, and (3) the limited time before the content goes into the commons. These three exceptions have been the standard protections for the public built into copyright legislation by Congress and the courts. First, the fair use doctrine gives the public a limited use of copyrighted material for educational and scientific purposes. Second, the first-sale doctrine stipulates that the copyright matters for the first sale of a commodity only; for instance, it allows people to resell CDs, both new and used. Third, fundamental to the U.S. Constitution is the notion of releasing intellectual property back to the public domain after a brief period; intellectual property was never intended to be a perpetual monopoly. All three of these exceptions were created because intellectual property was always supposed to remain a *public* good (Boyle 2008). Lessig contends that while these uses are not constrained by law, authors can still limit them by not allowing the public access to their work in the first place; however, once intellectual property becomes available to the public in a non-digital environment, it is impossible to restrict access to it. Through the fair use doctrine, courts and legislators have acknowledged that cultural production does not happen in a vacuum; rather, cultural goods are the product of previous cultural artifacts. Copyrights can restrict only very specific uses without the permission of the author because cultural goods are always already the products of previous creative work (Coombe 1998)—that is, a song may be written by a composer, but that composer was influenced by other composers. These interests are imperative for copyright legislation, but the public is largely excluded from negotiations.

Finally, musicians (or content creators more generally) are the group most severely impacted by not being at the copyright negotiation table because it is they who create the music that is ultimately copyrighted, owned, and commodified. Copyright legislation structures the relationship between musicians and record labels by specifying precisely how intellectual property can be bought, sold, commodified, and protected; musicians give up these rights when they sign a record contract. However, one group at the multiparty copyright negotiations professes to represent the needs of musicians: record labels (represented by the RIAA). Since record labels own copyrights and purchase those copyrights from the musicians that they have contracts with, they claim to be representatives of the musicians' interests. However, record labels have divergent interests from musicians.

There are two major problems with the record labels' position: (1) the vast majority of musicians, and by extension copyright owners, are independent of any record label (let alone a major record label); (2) it is precisely the relationship between major record labels and musicians via copyrights that creates exploitation in the recording industry. These two points highlight a contradiction that Marx pointed to in "Wage-Labour and Capital" (2000) and which has been increasingly a problem for labor in the United States since the 1970s: labor does not have the same interests as capital, and those interests are in fact diametrically opposed. Even though musicians earn their wages through the profits of record labels, it is precisely in the claims that the record labels make to advance the interests of musicians that the exploitation of those musicians' labor exists. When new copyright legislation is created with record labels as the representatives of musicians' interests, it is created in a way that places musicians against consumers and conceals the conflicting interests of record labels and musicians.

The multiparty copyright negotiation process runs counter to how liberal pluralists explain the way that a democracy works. For instance, Robert Dahl (2005) contends that, in a plural democracy, the role of the government is to mediate between groups of various interests. In a plural democracy, Dahl argues, different groups with competing interests get the government to enact legislation in their interest at various times. For Dahl, pluralism is based on the complexity of American society, which creates multiple sites of identification for its citizens. Since most citizens belong to more than one competing identity, alliances materialize a number of ways in society. Those alliances determine for whom one votes. The changing alliances allow different groups to govern at different times.

However, the premise of multiparty copyright negotiations is that there are limited groups that have an interest in copyright legislation. While the government mediates the demands of these groups, the narrow construction of who is interested in these negotiations leaves groups out of the negotiations. Rather than a liberal pluralist democracy, the multiparty copyright negotiation system resembles a system that allows corporate elites to construct the laws that others must follow. Not only are new industries, the public, and musicians excluded from the multiparty negotiation process, but their exclusion precludes them from having their interests considered; for instance, their exclusion from DMCA negotiations resulted in a law that benefits "major stakeholders at the expense of the public at large" and is "hostile to potential new competitors" (Litman 2006, 144). By not including these groups, the state negates the fact that the lives of people in these groups are affected by changes in the law.

Record labels, among other copyright stakeholders, recognize that the Internet and other forms of digital media have the potential to disturb the copyright balance of power in the music industry. To make sure that the copyright balance of power is maintained, copyright stakeholders initiated the negotiation process several times over the past two decades to alter copyright law. It is significant to note that copyright stakeholders initiated the process: not Congress and not the public. What this demonstrates is that through the rewriting of copyright law, the current copyright stakeholders ensure that they retain their power, copyright maintains its position over content, and content remains a capitalist commodity in a digital media regime.

LISTENING IN AN ANALOG WORLD

In 1995, the Internet was still in its infancy. The only way to purchase music was in a store, over the phone, or through the mail. Peer-to-peer (P2P) file-sharing programs had not yet been developed, and the iTunes store was nearly ten years from its launch. Music was firmly associated with physical objects. When consumers purchased music, there was no way of knowing what they actually did with their purchase. In this period, the primary media format that music was sold in was the CD, but tape cassettes and LP records were still available in stores. Physical home reproduction of music was possible by copying CDs onto tapes. This form of reproduction created barriers for large-scale reproduction: tapes are not particularly durable, and each subsequent copy of an original CD (i.e., each copy of a copy) produces music of decreasing sound quality. While the digital audio tape (DAT) created higher-quality replication, the Audio Home Recording Act (AHRA) of 1992 made the technology overly cumbersome and too expensive to make large-scale reproduction possible.

After purchasing a physical music recording, music consumers were permitted to do just about anything with the music. Someone who purchased a CD could listen to that CD as many times as he or she liked. If they wanted to play the album for friends, they were free to do so as long as the song was not being publicly performed.[7] This person could also lend the CD to a friend, and the friend could once again listen to the CD as many times as they wanted. There were no restrictions on who could listen to a CD, nor were there restrictions on who could listen to the CD independently of the original owner. Furthermore, the person who originally purchased the CD could resell it under the first-sale doctrine in copyright law. In turn, the copyright owners of the CD did not expect any royalties from the resale of the CD—these exchanges could happen indefinitely without any interaction with copyright owners.

Copying purchased music was in no way restricted as long as it was not for commercial sale. A person could buy a CD and make as many copies as they wanted for personal use. Furthermore, they could copy the entire CD, or just certain songs, and give it to friends. Mix tapes are a cultural form of exchange where fans can make their own varied compilations to share new music with friends. Beyond commercial sale, in 1995 the only restrictions on copying were technological restrictions placed on DAT recorders that created progressively degraded quality for copies of copies.

This is a description of the media environment that copyright laws were designed to regulate in the mid-1990s. As the Internet and digital technologies began to alter the production, distribution, and consumption of music, the copyright laws in place in the United States no longer addressed the demands of copyright stakeholders. Consequently, copyright stakeholders began to develop new copyright laws that address the relationship between the Internet and digital technologies and the consumption of music. While the media system described here has changed dramatically, the legal and policy framework that regulates music has shifted extensively as well.

Chapter Four

Critical Junctures

At moments of transformation in the means of production, capital attempts to modify the policies under which capitalism functions to ensure the highest rate of profit possible. In several books over the past decade, Robert McChesney has insisted that citizens must stand up at this critical juncture of media policy and demand rights. A critical juncture of media policy is a "period in which the old institutions and mores are collapsing under long-run and powerful pressures" (McChesney 2007b, 1434). At these moments, the way that we communicate is altered drastically by changes in technology, policy, and culture. Critical junctures are so important because the policies made with regard to media in these moments will be difficult to alter for decades (McChesney 2007a, 9). While much of the literature on the Internet's subversive qualities assumes that the Internet can only bring more democratic speech and press (Jenkins 2006; Rheingold 2000), McChesney stands out as a scholar who, while claiming that the Internet is not inherently democratic, is willing to become politically engaged to create more equitable media policies. By examining the importance of critical junctures of media policy on media regimes, I hope to demonstrate why it is so important for citizens to pay attention to changes of state policy with regard to music. My objective here is to lay out the rationale for calling the recent past a critical juncture of media policy, then to analyze whether the current moment remains a critical juncture.

The problem that McChesney articulates is that Congress usually creates media policy without input from the American public. This has been a substantial part of McChesney's work throughout his career, and he is one of the most thorough scholars at describing the way that media policy is created behind closed doors to benefit large corporations. McChesney's general argument is that the public must demand a seat at the stakeholder table in deciding communication policy because a functional public sphere, fostered by the media, is vital for a working democracy. Without a free and accessible public sphere democracy is impossible because citizens need access to information and a place where they can debate. The recording industry was so important during the digital transformation because music exposes people to ideas in what should be an open public sphere. Since the recorded music is controlled by a small oligopoly of record labels, the content in the public sphere is censored to meet the corporate interests of those labels.

McChesney lists three criteria that distinguish the existence of a critical juncture in media policy. He explains that "critical junctures in media and communication tend to occur when at least two of the following three conditions hold:

1. There is a revolutionary new communication technology that undermines the existing system;
2. The content of the media system, especially the journalism, is increasingly discredited and seen as illegitimate; and
3. There is a major political crisis in which the existing order is no longer working and there are major movements for social reform." (McChesney 2007b, 1434)

McChesney contends that two of the three conditions have been met. First, the Internet and digital communication technologies created a policy vacuum that the U.S. government was not prepared to regulate. Second, in 2003 and 2004, McChesney points out that some Americans were demonstrating increasing dissatisfaction with the media system by criticizing the concentration of media systems in the hands of a few. This dissatisfaction occurred as a result of the Federal Communications Commission's (FCC) policy changes to media ownership and culminated in the Supreme Court's overturning of the FCC's media ownership policy (McChesney 2007a).

Throughout the recent media transition, the music industry shifted its business model significantly because of the online availability of digital music. The dominant actor in the music industry is currently the recording industry. Digital music changed the political economy of the recording industry because the recording industry has begun to rely more on the exploitation of copyrights than the selling of physical media. Disintermediation—the elimination of intermediaries in the distribution process—created a disjuncture where consumption does not involve the material consumption of physical goods; people no longer need to buy commodities to receive musical content. Some people would counter that we do in fact need to pay for intellectual property even if we do not need the physical thing because intellectual property exists as a means to reward the creative process; creators would have no incentive to create without the protection of intellectual property. Whereas the Constitution had basic legal rationale about authorship and scientific progress related to copyright, this argument assumes that there is some rational legal basis behind the writing and implementation of copyright legislation. The problem is that a specific set of copyright policies are designed, not to protect authorship, but rather to privilege some interests over others; the current system forces recording artists to cede ownership to gain access to audiences. As discussed above, Jessica Litman demonstrates that while Americans tend to believe that intellectual property adheres to a principle (2006, 13), it is actually the product of negotiations between stakeholders, and these negotiations happen without input from the largest stakeholders of all: the public. In effect, the major record labels are doing more than changing their business practices; they are fundamentally changing the legal and technological structures upon which their businesses work.

As McChesney details, moments of transformation in the way we communicate are always met with changes in policy. While McChesney argues that the public must actively get involved in media policy, Litman contends that current copyright stakeholders continue to change copyright policies with multiparty negotiations. Over the past two and a half decades, the recording industry created a long paper trail of bills and laws to modify the system in which they operate.

LAWS CREATED AT THE RECENT JUNCTURE

Beginning in the early 1990s, the recording industry successfully lobbied Congress to pass a number of bills that strengthened the position of the major record labels in the broader music industry. At critical junctures, laws are passed that change the media regime for decades to come; thus, there is a need for the people to advocate for their interests in discussions about these policies. However, there have already been a number of government policy changes to copyright laws that happened without input from the people. Here I scrutinize three policies that alter media consumption to fit the transformation to digital media and directly affect the music industry. An analysis of these laws demonstrates that copyright policy has changed in profound ways that began even before the development of P2P file sharing. Instead of having a robust public debate about media policy at the recent critical juncture of media policy, the copyright stakeholders used their multiparty negotiations to create a new copyright apparatus before the public became aware of the crisis. These three laws demonstrate the power that major record labels have to shape the rules that regulate their industry.

First, in 1992, before the start of the current critical juncture, President George H. W. Bush signed the Audio Home Recording Act (AHRA) into law. The AHRA is significant because it is the first law to address digital audio reproduction. The problem that the recording industry sought to address with the AHRA was that digital copying on a digital audio tape (DAT) allows for an infinite number of perfect copies. Since digital players were equipped with a record button, the recording industry wanted to make DAT players illegal; if major record labels could not block the players, they threatened to not issue licenses for DAT music. The AHRA was an agreement between current stakeholders, copyright owners, and consumer electronics manufacturers where copyright owners agreed to make digital players legal in exchange for royalties "on every digital recording device player and digital tape sold" (Litman 2006, 61). Furthermore, while the AHRA permitted noncommercial home taping, it prohibits the copying of copies through a technological fix. The AHRA requires that DAT machines be made to recognize computer code that would degrade the quality of each respective copy to make DATs more similar to tape cassettes; this is the first instance of digital rights management (DRM)—computer code that automatically restricts the use of digital content—in legislation in the United States. DAT players are permitted to make copies of the original recording at the highest fidelity, but when the code on the digital recording identifies a recording as a copy of a copy, it records the music at a progressively lower fidelity for each consecutive recording (Lessig 2006). This marks the first time that consumers lost the ability to do what they wanted with a music commodity, and they had no say in the law. Restricting the reproduction of music recordings in the AHRA is one of the first attempts by Congress, in collaboration with the recording industry, to restrict the manner in which consumers listen to music.

Furthermore, copyright owners secured provisions in the AHRA that require royalties to be paid for blank digital media. These provisions of the AHRA that award royalties to record labels are perplexing because they are a state intervention on behalf of major record labels that assume that consumers purchase blank digital media with the intention to violate copyright law, and the AHRA maintains a right on behalf of the major record labels to revenue from this consumption. These royalties apply to DAT recorders, blank CDs labeled for music, and stand-alone CD burners (Hall 2002), regardless of whether or not they were used to record copyrighted material. Blank media have no intentional purpose: when someone buys a blank DAT, there is no way to tell whether that person will use the tape at all, record copyrighted material, record his or her own music (or other material), or record music with permission from a copyright holder. Yet the law assumes that people will violate copyright law with every purchase of a digital tape.

Furthermore, the AHRA assumes that when people violate copyright law, they are violating the copyright of the major record labels. While the law does not pay the royalties directly to the major record labels, it pays them to the Alliance of Artists and Recording Companies (AARC), which primarily pays revenue to the major record labels and the biggest artists on those labels; performance rights organizations (PROs) distribute the rest of the proceeds to artists. In this way, royalties are distributed based on the same system that royalties are distributed for music playing in a bar; PROs determine the distribution of royalties based on radio airplay. These royalties are not allocated to musicians based on the actual violation of their copyrights. This does not account for music fans who copy noncopyrighted music or music by smaller artists/bands or people who are backing up computer files using digital media originally intended for music. The AHRA establishes the precedent that all digital copying is a violation of the copyrights owned by major record labels without regard to the actual copying practices of individuals.

Second, President Bill Clinton signed the Digital Performance Right in Sound Recordings Act (DPRA) into law in 1995 to create a performance right for music distributed through digital means (Holland 1995). This legislation demonstrates that in 1995[1] the music industry was already taking note of the potential impact of digital music on the Internet. While the MP3 was not released until 1995, streaming Internet radio was broadly popular by 1994 (Boehlert 1994). By 1995, the free download for Real Player was giving all computers access to Internet radio broadcasts. Internet connections were fast enough in 1995 to get low-quality music to users, and it was becoming apparent that media consumption was going to be a big part of Internet usage (Jeffrey 1997). One thing that the recording industry feared was that, without regulation of music broadcasting on the Internet, people could begin to post entire albums on the Internet and broadcast them to others (Atwood 1995; Block 1995). Streaming music without downloading it to a hard drive could interrupt profits for the major record labels. Recognizing the potential impact of Internet radio on their business model (Boliek 1997), the Recording Industry Association of America (RIAA) lobbied for the DPRA on behalf of major record labels.

On its most basic level, the DPRA creates a performance right on Internet broadcasts that never existed for analog radio. One massive loophole in the 1976 Copyright Act exempted broadcast radio from paying performance rights fees to recording artists: only composers/authors receive royalties for the airplay of a song.[2] This is partly a result of the system of payola. Record labels and recording artists have always tried to induce radio stations to spin albums and singles because radio airplay is a type of promotion. Payola is a system where record labels pay stations to play their artists' songs. Having music played widely is vitally important for record labels to promote their music, and payola is often seen as no different from advertising; however, the payola system is not the same as advertising. Radio stations play the music as the product of their business model instead of as an advertisement. This creates a strange situation because copyright law stipulates that copyright holders must be compensated for the public performance of their work.

Moreover, since record labels pay radio stations to play music under the payola system, labels would in effect be paying themselves. Exempting radio broadcasts from having to pay certain performance rights helped to foster payola, but it also demonstrates concessions that were made during previous multiparty negotiations of copyright legislation.

Radio was at one time the dominant industry within the broader music industry (1930s–1940s). Having these royalties waived in the current iteration of copyright signifies radio's previous power. This created a conflict between the RIAA and the National Association of Broadcasters (NAB) because the NAB was unwilling to make concessions over paying for performance rights in any format (Saxe 2001b). Internet radio was a new stakeholder at the passage of the DPRA and therefore was excluded from the negotiation table; however, analog broadcasters fought the DPRA because many analog radio stations already had plans to simulcast over the Internet (Saxe 2001a). Creation of the DPRA demonstrates not only that the recording industry was aware of the potential impact of Internet distribution, but also that there are constant internal power struggles among parts of the broader music industry.

The third, and possibly most significant, piece of legislation passed during the recent critical juncture was the Digital Millennium Copyright Act (DMCA) of 1998.[3] In 1995, a White Paper, that would later become the DMCA became the first official legal analysis to assert that every time someone executes a file on a computer, it is a reproduction because it is "copied" in random-access memory (RAM). Defining every copy in RAM as a reproduction that requires authorization by a copyright owner would have effectively shut down digital content, so this gave content owners an advantage at the copyright bargaining table because by making this argument, the recording industry began the discussion with copyright law already working in its favor. As discussed above, copyright stakeholders often characterize current law in a way that demonstrates that they already have the law on their side before arriving at the multiparty negotiation table (Litman 2006). Prior to the interpretation of all RAM copies as a reproduction, copying involved a certain amount of intent, whereas executing a music file on a computer more closely resembles, to consumers, playing an already purchased CD, tape cassette, or LP. By arguing that RAM copies are unauthorized copies, the recording industry fundamentally changed what it means to consume music because all digital listening requires making copies of a file. Fortunately, the DMCA passed without this extreme position, but the recording industry used it to get a more favorable deal, through compromise, at the multiparty negotiations to update the copyright law.

While the DMCA does not outlaw executing music files on computers, it does fundamentally change the way people consume music by allowing content owners to restrict the way that music listeners can *use* music. The

DMCA, following the AHRA, encourages the use of digital rights management to limit unauthorized copies of content. As an anti-circumvention technology, DRM adds "code to digital content that disables the simple ability to copy or distribute that content" (Lessig 2006, 116). However, the DMCA does not mandate DRM, but rather makes circumventing DRM illegal. By criminalizing DRM circumvention, the government made it impossible for consumers to use their music in any way that goes against the rules desired by copyright owners.

Under the DMCA, digital copyright law shifted from the prohibition of copies for commercial profit to the regulation of consumption because DRM allows content owners to place restrictions on the way music consumers use music. First, DRM can restrict the number of computers on which a file can be stored. Second, it can further restrict the number of times a file can be used and in what ways it can be used. Third, DRM can provide information about how the music is being used to content owners. This is a redefinition of the traditional role that copyright, and intellectual property in general, plays in music. Record labels are seeking new ways to control and enforce the consumption of music.

While the DMCA was created through multiparty negotiations, the bill took a quite different path to passage than the AHRA and the DPRA. Upon the release of the White Paper in 1995, the Digital Future Coalition[4] and the Home Recording Rights Coalition[5] implemented lobbying campaigns to block the passage of the DMCA (Litman 2006, 124). The lobbying effort forced the copyright stakeholders to go back to multiparty negotiations and form another strategy. Bruce Lehman, commissioner of the United States Patent and Trademark Office under President Clinton, decided to try to get the DMCA attached to a treaty through the World Intellectual Property Organization (WIPO), since member countries of the WIPO are forced to adopt legislation to implement its treaties (Litman 2006, 129). Ultimately, this approach worked. The DMCA passed both houses of Congress, and President Clinton signed the law under the line of reasoning that the United States must honor its treaty obligations. This raises a major question about democracy: when widespread opposition to a bill blocks its passage, what does it mean when that bill ultimately passes, without discussion, to make U.S. law compatible with an international treaty? While international organizations may represent the will of the international community,[6] they do not represent the interests of the public in the individual nation-states. The power of multinational corporations to pass legislation in their own interest, through whatever means necessary, demonstrates the power of capital over the interests of the people, as when, despite opposition to the DMCA, content industries were able to circumvent democratic practices and force its passage.

During the multiparty negotiations for each of these three laws, the recording industry was in conflict over copyrights with a variety of different stakeholders, but the negotiations themselves never considered the interests of musicians or the public at large. Negotiations for the AHRA were primarily between the recording industry and the consumer electronics manufacturers, as the consumer electronics manufacturers wanted to produce and sell new devices that would allow their users to produce near-perfect copies (Litman 2006, 59–61). When the DPRA was created, the multiparty negotiations were aimed at reconciling the interests of the recording industry, broadcasters, and PROs. The resulting law of these negotiations is particularly detailed in its exemptions to meet the interests of industries at the table. The third set of negotiations focused around reconciling the demands of the recording industry, the telecommunications industry (specifically Internet service providers), and consumer electronics manufacturers. None of the negotiations for these changes was based clearly on any root copyright principles, and the resultant legislation is complicated due to the unique demands of different stakeholders; however, the state felt compelled to act, and it acted in favor of corporate interests. Record labels were the political winners at the copyright negotiation table during the digital transformation because the AHRA established DRM and royalties for digital media, the DPRA gave the recording industry new sources of performance rights revenue, and the DMCA protected broader uses of DRM.

THE NET EFFECT ON COPYRIGHT

While copyright used to deal with the control over a song's reproduction and performance, the AHRA, DPRA, and DMCA have caused the *use* of songs to fall under copyright's purview. When the Internet first became a distribution channel for music, the record labels' stated concern was about widespread copyright infringement; in fact, the alterations to copyright mechanisms have done little to stop copyright infringement, but rather made it easier for copyright owners to restrict the ways that consumers use and consume music. These changes to the law reduce the autonomy of individual consumers to use the music commodities that they legally purchased. Even though the state has not fundamentally changed copyright, it has helped copyright owners by protecting DRM, which allows copyright owners to restrict the use of copyrighted material. While consumers are losing the ability to own music commodities fully, content owners are developing new ways to regulate the use of their intellectual property.

When music fans purchase tape cassettes, CDs, and LPs, they own a piece of *physical* property, but this significantly shifts with digital music under the new copyright regime. Copyright stipulates that consumers purchasing physical media can use that music in any way except reproducing it for profit or performing it publicly: they can play their music an infinite number of times;

they can give their property away as a gift; they can resell the CD, tape cassette, or LP; and, they can have their music stolen. However, digital music files are more limiting because their code (i.e., DRM) defines the exact ways that music can be used. In effect, DRM eliminates the public uses previously protected under copyright legislation: fair use becomes obsolete because computer code does not permit music listeners the ability to use it in a myriad of ways, and the first-sale doctrine no longer applies because digital music cannot be resold as "used" without making a copy.[7] As a result, it seems unnecessary to mention that someone cannot legally give digital music files to their friends. When we buy digital music, we already assume that it is illegal to share those files, but the naturalization of this process is partially hidden by code. Additionally, there is no mechanism in place to report stolen computer files.[8] The digital form of the music commodity fundamentally changes the way that consumers use their music by limiting their ability to share music and stopping them from playing it on an unlimited number of devices.

While the recording industry was aware of the potential of the Internet to change the balance of power in the music industry as early as 1993,[9] the change that came as a result of the MP3 proved a direct threat to the recording industry's business model. There were early attempts by the recording industry to create protected digital files, but these attempts were not commercially successful; during the mid-1990s, the recording industry experimented with Liquid Audio and a2b, but these formats "were incompatible and clumsy, and consumers seemed generally uninterested" (Gillespie 2007, 42). In late 1998 (following the release of Napster), the recording industry, consumer electronics manufacturers, and other copyright stakeholders got together to try to create their own files. This forum came to be known as the Secure Digital Music Initiative (SDMI). Record labels, however, had a difficult time encouraging music fans to use the resulting SDMI file type because MP3s were already so prevalent. The problem for the recording industry's business model was the unsecure nature of the MP3 file; computer users could easily copy music from CDs onto their computers and share them with other computer users over the Internet. P2P software, like Napster, Kazaa, and Gnutella, created a digital environment where P2P users could easily send files over the Internet to other users. MP3 files appeared to the recording industry to be dramatically reducing the need for music fans to purchase recorded music and to be threatening the recording industry's profit margins.

Since the recording industry appeared to be having such a difficult time competing with MP3 files in the late 1990s and early 2000s, its initial reaction was to shut down access to MP3 files. The RIAA filed a lawsuit against Napster in 1999 that ultimately shut down the P2P service in 2001 (Napster subsequently reopened as a subscription service). Then from 2003 to 2008, the RIAA attempted to deter file sharing by pursuing individual users through lawsuits and, in effect, criminalizing the sharing of music. Currently, the RIAA uses ISPs to issue cease-and-desist orders to file sharers instead of suing them (Bruno 2009).

However, all of these methods are reactionary rather than preemptive because they aim to punish people for purportedly violating copyright instead of preventing them from violating it. To prevent file sharing, the RIAA began to develop DRM to restrict the ways that a given file can be used (Park 2007), along the way avoiding the question of the legality of file sharing. While the AHRA created the first computer code to restrict the digital reproduction of DAT music, the RIAA's new model of DRM would be highly restrictive (Lessig 2006, 116). Advanced Audio Coding (AAC) files were the first widely used DRM-encoded music files when Apple began using them on iTunes in 2003. AAC files not only prevent copyright violations but also restrict the way that consumers can use their music (Park 2007). By using DRM to prevent the unauthorized reproduction of music, the recording industry not only produces a technological fix for alleged piracy; they also prevent music fans from participating in previously legitimate listening practices, reducing the autonomy of music listeners.

The shift from restricting unauthorized performance and reproduction to restricting use was not brought about by any fundamental change in copyright law but rather by a new protection granted to computer code that had previously been unnecessary. As Lessig argues, code has become the law (Lessig 2004; Lessig 2006), that is, computer code regulates behavior. Computer code attached to music files acts to regulate/restrict the behavior of music listeners. In *Code: Version 2.0*, Lessig argues that behavior can be regulated through a number of mechanisms, including "the law, social norms, the market and architecture" (2006, 123). DRM is the architecture that now stipulates the ways digital music files can be used. The law does not define what exactly content owners can protect with DRM; rather, it stipulates the protection of this code through the DMCA because the DMCA criminalizes the circumvention of DRM. While copyright is still protected by law and copyright owners are still willing to prosecute copyright violators, code is beginning to preempt people's ability to test copyright law through the judicial system.

As the system that makes it difficult for authors to restrict the use of their content is replaced by a digital system where they can exercise near-perfect control over their creations, many content owners do not provide room for public/fair uses in their code. Lessig describes the way that computer code can restrict access to digital files in ways that the law was never designed to restrict; in turn, "code becomes law" (Lessig 2004, 100). Tarleton Gillespie asserts that while content owners create a debate among the public over piracy, they have already implemented policies that concretize the content owner's position. "At the core of these changes is a fundamental shift in strategy," Gillespie maintains, "from regulating the use of technology through law to regulating the design of the technology so as to constrain use" (2007, 6). While intellectual property law was created to *regulate* how individuals *reproduce* copyrighted material, the new policies *restrict* the actual *use* of content.

Incorporating DRM into music files enables a digital enclosure (Andrejevic 2007) that goes well beyond the initial property enclosure created by copyrights because it eliminates a considerable part of public use. Even after copyright was established, the public's interests were protected by the fair use doctrine because courts and legislators acknowledged that there were circumstances when the public good outweighed the interests of the author. One risk of "an unlimited copyright is that copyright owners might be able to enforce their rights in ways that inhibit the production of new work, squelch the expression of particular opinions, or undermine the health of public discourse" (Gillespie 2007, 29). Enabling broad uses of copyrighted work for the public fosters free speech protected by the First Amendment. Under the nondigital copyright regime in the United States, this meant that people have to request permission to use copyrighted materials only under certain circumstances (e.g., recording a new version of a song written by someone else). The digital enclosure of copyright changes all of this by denying the public access to certain uses of a copyrighted work through code. DRM does not allow the public to use digital copyrighted material without previous permission. Since "part of the premise of fair use was that such uses should not require asking permission of the owner" (Gillespie 2007, 63), allowing DRM to restrict use is a dramatic change in the application of copyright law. Digital music requires people to ask permission of copyright users before they can use copyrighted material in a fair way; this effectively eliminates the fair use doctrine because copyright owners' organizational structure does not allow them to process fair use requests.

Regulating music consumption by computer code has already fundamentally changed the way that music consumers can listen to their legally purchased music. The first way that DRM restricts music consumption is by limiting the number of computers on which a file can be stored. iTunes' protected AAC files (also known as M4P) stipulated that the files could only be stored on five computers.[10] While five computers may have seemed sufficient in 2003, it is quite conceivable that a household can now have more than five computers/devices (e.g., iPod, tablet computer, smartphone, Xbox, laptop, desktop, etc.); therefore, not every computer that a person owns could necessarily store all of that person's music collection.[11] The M4P also restricted the number of CDs that a file could be burnt onto as part of a playlist. In the case of iTunes, the most egregious file restriction was that M4P files could be played only on Apple software and hardware, so if a person legally bought music from the iTunes store and attempted to play it on their Zune player, the music file would not work. People who consume music via a subscription service are in an even more tenuous relationship to the music they consume because subscription music contains DRM that can sense whether a subscriber is still subscribed to the service. When files downloaded from a subscription service are connected to the Internet, they automatically check a user's subscription status; if they detect that the person is no longer a subscriber, the files begin to lose their functionality. Former subscribers lose their music libraries because they are not paying to own music; if they want the music that they have been listening to, they must once again pay for that music. Finally, there is no way to give DRM music as a gift outside of giving a gift card because digital files cannot be transferred in the same way as a CD. None of these changes has been the consequence of modifications to the law, but rather these changes are a result of computer code.

Changes in copyright laws during the 1990s have had concrete effects on the way that music is listened to and consumed today. This was the result of active engagement on the part of the copyright stakeholders at multiparty negotiations, which entirely left out the needs of the public. The question remains whether citizens still have the power to change U.S. policy in a substantial way to create more use that is public and gain more autonomy over their listening practices.

A CRITICAL JUNCTURE PASSED?

At moments of critical juncture in media policy, the laws passed impact the media environment for decades to come. McChesney advocates that people to demand action from their elected officials to foster a more robust free public sphere at these moments. One shortcoming in McChesney's argument is an issue of timing. In *The Problem of the Media* (2004), McChesney is adamant that a veritable uprising happened in 2003. He argues that in 2003 people were beginning to feel concerned that too few corporations owned too many communication channels, and he uses polls to demonstrate this. McChesney also points to the willingness of politicians to make this a campaign issue for the 2004 elections. More importantly, news media were actually beginning to cover this story; McChesney thinks this might turn into active public discussion of and participation in the making of media policy. However, as the three laws outlined above demonstrate, content owners were already hard at work reinforcing their position in media production long before 2003–2004. Ignoring these concrete policy changes, McChesney became hopeful that with public support network neutrality could be implemented to maintain the Internet as a neutral space in terms of what is being transmitted over it. However, I argue that the state already passed laws that significantly alter the media system for the benefit of copyright owners, and these laws, along with Internet user habits, will shape the media environment for the near future.

Network neutrality (hereafter net neutrality) was supposed to be a reform on the side of the public sphere that helps maintain the "open" Internet. At its core, net neutrality is a principle "by which faraway ISPs would not be permitted to come between external content or a service provider and their customers" (Zittrain 2008, 178). In other words, there has been a growing fear that Internet service providers (ISPs) will restrict bandwidth, or even access altogether, to certain sites; ISPs could even charge websites to have access to larger bandwidth. As a result, net neutrality is an important issue that will have a substantial effect on media distribution. While some websites would benefit from paying a fee to ISPs, the vast majority of Internet start-ups and small-business websites would be unable to get content to consumers in an efficient way. This is especially harmful to independent content producers because they will not have the capital to get their products to their fans. McChesney and net neutrality advocates argue that an open Internet with automatic net neutrality would allow Internet users to access information that is more diverse. In the music industry, there are advocates who argue that the Internet creates a space where independent musicians can compete with musicians signed to major record labels.[12] The argument goes that under a physical media regime, independent musicians cannot compete with the major record labels; however, on the Internet, any musician can start a website (or use social networking sites like MySpace or Facebook) and begin to distribute music. Allowing ISPs to restrict bandwidth by charging fees to websites would decimate the ability of independent musicians to distribute music on the Internet; however, the Internet already presents barriers for independent musicians to distribute music online.

McChesney argues that the crescendo of public involvement in media policy came in a debate about net neutrality. The public became concerned that ISPs would begin to sell access to bandwidth. Arguably, this would limit the ability of Americans to access websites freely and would funnel Internet traffic to a few wealthy and well-established websites. While McChesney saw this becoming an issue in 2003–2004, it continued to grow into 2010. The public was handed a huge victory on this front on December 21, 2010, when the FCC, under the leadership of Chairman Julius Genachowski, passed a policy for net neutrality. However, a federal appeals court overturned this in January 2014 (Fung 2014), placing net neutrality back in jeopardy.

While net neutrality could provide people with greater access to websites, Internet users have established ways of using the Internet that focus on a few media conglomerates. For news especially, people already visit a limited number of websites (Baker 2007, 113). Where there used to be an abundance of local newspapers around the United States, these newspapers have been closing because of readers dropping physical delivery subscriptions to read their news on the Internet. In 2011, the Project for Excellence in Journalism reported that 41 percent of news consumers obtained "most" of their news from the Internet (Rosenstiel and Mitchell 2011). As these news consumers have dropped their subscriptions, they have turned to fewer and fewer news sources; the readerships of the online versions of CNN, the *New York Times*, the *Washington Post*, Fox News, and the Huffington Post have skyrocketed (Rosenstiel 2009; Rosenstiel and Mitchell 2011). Edwin Baker explains that even though the Internet expands access to different types and sources of information, the "tendency toward concentration is, if anything, more powerful than in respect to offline media. The American public as a whole appears to be receiving more of its information from fewer sources" (2007, 113). Net neutrality could allow news consumers to find more news sources if they are persistent, but this does not change the fact that a large number of people are getting their news from fewer sources.

Furthermore, people have already established their music consumption habits online. It is interesting to note that the year of McChesney's public uprising, 2003, was the year that both iTunes launched and the year the RIAA commenced lawsuits against individual file sharers. These lawsuits were based largely on copyright legislation that was already altered during the recent critical juncture. Music consumers are inundated with a broad range of music on the Internet, but most people get their music from a few large websites. While some had hoped that the Internet would allow greater access to independent music because of the lack of intermediaries (Alderman 2002; Burkart and McCourt 2006), music consumers have flocked to large online retailers—iTunes is already the biggest music retail store ever (Christman 2010). While there are alternative online music retail sites, these sites are subject to conglomeration online as larger sites buy them; for instance, Amie Street—a site that specialized in independent music available at varying prices depending on popularity—was purchased by Amazon in 2010 (Associated Press 2010). Regardless of the FCC's net neutrality policy, music consumers already have established consumption habits and purchase music within a media environment established by laws passed during the recent critical juncture.

The critical juncture of media policy that resulted from the development of the Internet and digital communication technologies has passed. People have developed media consumption patterns on the Internet that take them to a few major conglomerates for most of their consumption. While Congress could still implement laws that influence Internet usage, the media environment on the Internet is already established. In the music industry, the AHRA, DPRA, and DMCA collectively establish a media environment that employs DRM to restrict the use of digital copyrighted content. This media environment establishes the parameters under which media will operate for decades.

CONCLUSION

The digital transformation of the music commodity presented a moment when copyright could no longer maintain the recording industry's dominant position in the music industry. However, due to the diligent work of recording industry representatives, copyright was changed to accommodate the wishes of major record labels. This process began before P2P file sharing became an issue, so the recording industry was actively engaged in restructuring the legal system before the time when it claims Napster surprised it. As copyright laws become antiquated, the state provides the means through which major record labels can restructure consumer behavior. There was no underlying principle that guided the enactment of the AHRA, DPRA, and DMCA, other than the maintaining of the current balance of power among copyright stakeholders. The U.S. government fundamentally altered copyright because these laws collectively changed access to and use of copyrighted material in digital formats by developing DRM. While more than a decade has passed since the passage of the DMCA, there has been no real attempt to pass legislation that would dilute copyrights in the digital era; furthermore, court cases and regulatory bodies have strengthened copyrights for rights holders. Changes in copyright law also change the relationship between capital (record labels) and labor (musicians); the next part of *iTake-Over* will follow the ways that this relationship has changed because of the digital transformation of music.

Part III

The Recording Industry and Labor

Chapter Five

Musician Labor

Central to the piracy panic narrative is the claim that file sharing hurts musicians. But how is the labor of musicians exploited? In what way did the social relations of production change during the digital transformation? How does the piracy panic narrative obscure the relations of production in the recording industry? Musicians function as labor in the recording industry, and while most commentators tend to cast musicians as "artists," the appellation only obfuscates their role as laborers and their work as labor.

As the recording industry adjusted its means of production to digital technologies, the relations of production between capital and labor faced corresponding changes. Unfortunately, theories of the labor of musicians as the primary workers in the recording industry are rare. Since musicians often are labeled as artists, their labor tends to be disregarded as inherently different from that of other workers; this is the same barrier that has been erected repeatedly between knowledge workers and physical workers (Schiller 1996). In fact, the term *recording artist* suggests an artistic autonomy that record contracts specifically take away. The failure to understand critically the status of musicians as workers has created an environment in which those very musicians are expected to undercut the value of their labor power for the success of the recording industry.

By failing to identify musicians as workers, the RIAA constructs musicians as victims of their own consumers. The bulk of this error in perception stems from the construction of musicians as "artists." This construction seems to happen for two reasons. First, the perception of art as somehow outside of the capitalist system persists in contemporary culture. Second, record label contracts position the musicians as "recording artists" (or simply as "artists"). Arguably, the most important division of a record label is the artists and repertoire (A&R) division, in which label employees decide which musicians to sign and what type of music the label should attempt to produce and sell. The status of "artists," assigned to musicians through record contracts, requires further interrogation; the following is the beginning of such an interrogation, through which I examine the effects that the digital transformation had on musician labor.

CREATIVE PRODUCTION

While the ideology of the musician as artist obscures the material labor that musicians produce as workers, it is equally important to understand how the "artist" has functioned as a site of creative production. Cultural studies provides Marxism with an important contribution to this thought because scholars working in the cultural studies tradition want to understand culture in a broader sense, beyond political economy, in order to understand other benefits that musicians derive from their creative work. On the other hand, workers involved in so-called creative production are often fetishized by other workers because of their ability to get enjoyment out of their work (Schiller 1996). Since creative workers derive enjoyment out of their labor, other types of workers sometimes argue that the production of creative workers remains autonomous and outside of the logic of capitalism—Dan Schiller (1996) describes and takes issue with this position. However, it is exactly this ideology of creative production that allows for the material exploitation of creative workers.

The problem with the ideology of the artist is that it assumes that creative production somehow transcends capitalism. In *The Field of Cultural Production* (1993), Pierre Bourdieu uses his framework on "field" and *habitus* developed in *Distinction* (1984) to describe the ways cultural production adheres to a capitalist logic, even when cultural products appear to constitute an artisanal or alternative production model. A field is "a separate social universe having its own laws of functioning independent of those of politics and the economy" (Bourdieu 1993, 162). In any given field, people are classed, and their position within the field influences the way they interpret the world and interact with others within the field. A field consists of a dominant class and a dominated class, and within each class, there is both a dominant fraction and a dominated fraction (Bourdieu 1984). While people may be located in a specific class and a specific fraction, Bourdieu argues that they can move between classes and class fractions; however, it is easier for people to move in certain directions (i.e., across class fractions) than in other directions (i.e., from the dominated fraction of the dominated class to the dominant fraction of the dominant class). People's *habitus*—the structuring dispositions that a person possesses—allows them to negotiate the field in a way that does not predetermine outcomes. Within a given field, possession of and access to different types of capital (e.g., economic, cultural, or social capital) determines the location of a person within the field, but that person's *habitus* helps to determine where they can move within the larger field. Since cultural production is often described as transcending the logic of capitalism,[1] Bourdieu's work is important because it reasserts the role that capital plays in situating cultural producers within capitalism.

Capitalism operates in cultural production even when the relations of production do not involve capital buying labor power from workers. Rather, Bourdieu claims that capitalism is located in the class position of cultural producers in the field of cultural production. Within the field of cultural production, Bourdieu contends that cultural producers tend to be in the dominated fraction of the dominant class, while their work is situated in society in different class positions. Bourdieu stresses, "The struggle in the field of cultural production over the imposition of the legitimate mode of cultural production is inseparable from the struggle within the dominant class (with the opposition between 'artists' and 'bourgeois') to impose the dominant principle of domination" (1993, 41). The site of artistic struggle thus tends to be a struggle between fractions of the dominant class. An artist's autonomy from the system of capital is dependent on the artist's access to capital; this is especially true for artists who have little economic capital, as they must work at the service of capital (i.e., create mass culture), whereas a wealthy artist can work more autonomously (Bourdieu 1993, 67–68). Remarkably, this reverses the rhetoric of cultural struggle. For instance, an artist who rejects "selling out" and capitalism in order to create avant-garde art has to be in a relative position of dominance and affluence within the field of cultural production because he or she needs to possess economic capital in order to survive. On the other hand, an artist who lacks economic capital will be more willing to produce mass art in order to make ends meet. These two kinds of artists end up playing contradictory roles in terms of their status within the class field. Part of the trick to avant-garde art, however, is that the initial rejection of capitalism allows avant-garde material to become more valuable over time.

Bourdieu argues that cultural intermediaries within the field of cultural production help to create value for particular works of art (fine art, music, literature, etc.). By cultural intermediaries, Bourdieu means people such as art professors, music critics, museum curators, gallery owners, and the like. These cultural intermediaries create value by using their educational and cultural capital to define worthy art. In short, there is a degree of value created by critics and curators defining "good" art; these individuals build capital within certain circles or communities, and their opinion becomes the selling point. While Bourdieu focuses on cultural intermediaries in art, in the case of music, artists and repertoire (A&R) workers act as cultural intermediaries to define what type of music should be recorded and how that music should sound. As Bourdieu explains, "the professional ideology of producers-for-producers and their spokespeople establishes an opposition between creative liberty and the laws of the market, between works which create their public and works created by their public" (1993, 127). Cultural intermediaries are responsible for promoting art to the public, but in that promotion they help to construct a public.[2] While record labels ostensibly strive to meet the demands of a music market, A&R workers create a market for the music they want to sell. Recording artists have been valued for their supposed autonomy, but only to the degree to which supposed autonomy sells records.

Furthermore, the ideology of the artist is something that emerges under capitalism as a way to fetishize and alienate their labor. In *Bootlegging: Romanticism and Copyright in the Music Industry* (2005), Lee Marshall links the development of the ideology of the artist to Romanticism. For Marshall, the ideology of the artist and the construction of copyrights "provide a way of managing the commodification of culture in capitalist modernity" (2005, 2). This ideology is supposed to insulate the artist from the demands of a capitalist market, but rather it sustains the capitalist logic of the commodification of music. By seeing the ideology of the artist as linked directly to the capitalist demands of artistic production, it becomes clear that artists are no different from other workers.

While artists are always already workers that function to create commodities, some scholars do not see this connection as problematic. Among writers who argue that the autonomy of creative workers powers the economy, Richard Florida has taken on a predominant role in that cities hire him to learn how better to exploit creative production to increase GDP. Florida claims that creative forces are the strongest and most beneficial aspects of an economy today. In *The Rise of the Creative Class* (2004), Florida classifies those people who produce creative works as the "creative class." For him, these individuals are what drive the contemporary American economy, and without them the United States would lose its competitive edge.

Although this book is mainly aimed at explaining *why* we should consider these people a class based on *how* they interact, Florida does spend some time discussing the labor conditions he feels characterize the "creative" workplace. Florida argues that the economy is based increasingly on an educated population that has a difficult time conforming to the nine-to-five, suit-wearing culture of Fordist capitalism. Instead of restricting these workers, Florida says, companies need to embrace their creativity and allow them to work in whatever way they find comfortable. Florida thinks that the flexibility of the creative worker is creating a more humane work environment where workers ultimately have power because the worker owns the means of production in his or her head; therefore, creative work cannot be Taylorized. However, what is missing from Florida's argument is that increasing productivity has the effect of increasing the exploitation of each individual worker.

Whereas Florida sees great potential in this new creative class, Andrew Ross (2003) suspects that the new "flexible" workplace has the potential to develop new manifestations of labor's oppression by capital. In *No-Collar* (2003), Ross argues that while the "no-collar" work environment appears to be the archetype of humane work, it ends up working the laborer for longer hours and divorcing leisure from work entirely. Recording artists are the epitome of this "no-collar" work for theorists such as Florida because they can work when they please, wear whatever they want, and have the appearance of answering no one. For many of the new "creative" workers associated with computer software production, rock stars are the model of working upon which they base their work habits.

However, musicians sell their labor under the same relations of production as most other workers. Rather than idealizing the labor of musicians, it is important to look at how labor operates in the recording industry.

ACTUAL MUSICIAN LABOR

The recording industry distinguishes between recording artists and the studio or session musicians who perform the recorded music in the background of the named artist. Contractually, recording artists are independent businesses contracting with record labels under deals that stipulate the existence of the two as separate entities. On the other hand, studio musicians are paid as wageworkers to complete a particular task. Both sets of musicians, recording artists and studio musicians, play important creative roles within the studio, but their relationship to the label, and therefore the way that they are paid, differs both materially and rhetorically.

Materially, contracts bind both studio musicians and recording artists to execute certain duties in exchange for different forms of payment. In the case of studio musicians, contracts stipulate that they record a specific part for a song, for which they generally receive an hourly wage. These session gigs are highly coveted by musicians, and many musicians move to major recording cities (New York, Nashville, and Los Angeles) to get one of these coveted positions (Scott 2000). Producers have a list of the most in-demand musicians in a given city. These producers contact the most in-demand musician first, but if that musician is unavailable to record a session at that time, producers contact the next most in-demand studio musicians. Upon recording the album, the studio musician is paid the contracted amount, although on rare occasions he or she may receive a percentage of the royalties for albums sold. However, they are rarely employees of the labels, meaning that finding work, while steady, remains contingent; without being employees of the label, they do not receive full-time benefits (e.g., health insurance) and can never guarantee their next gig.

On the other hand, recording artists must meet an entirely different set of terms in their contracts. Rather than being in a wage relationship with the label, recording artists are contractually bound to produce a specific number of "sides" (i.e., songs) that are released on a number of albums (Hull, Hutchison, and Strasser 2011). Recording artists receive an advance from the label to record/produce, manufacture, and promote the album. In return for the advance, technical expertise, and distribution of their music, recording artists surrender copyrights and revenue to their label. Recording artists collect royalties on an album at a rate of 9 to 12 percent (higher for superstars) of the wholesale price of the album (Krasilovsky et al. 2007, 19; Park 2007). Since labels typically purchase all copyrights from their artists, royalties are the amount of money that labels agree to pay their artists for each sale of an album or song. However, before recording artists ever receive their royalties, they must first repay the original advance to their record label; at the same time that artists are repaying record labels for the advance, the record labels earn profit on the album.[3] Artists must recoup all money from the advance before they begin to earn money from royalties. For this reason, one way that the relationship between capital and labor often is described in the recording industry is that it is like that of a "share-cropper" (Slichter 2004) because artists are always paying off the debt that they owe their label. However, the debt does not follow artists outside of recording music. While these advances are different from a loan from a bank because they do not have to be paid back outside of the sale of albums, recording future albums is usually contingent on selling the first album well enough to come close to repaying the advance (Hull, Hutchison, and Strasser 2011; Krasilovsky et al. 2007). Most importantly, recording artists receive payment only after they have recouped the original advance; no matter how much labor recording artists put into selling the album, they receive their royalties only after the record labels have paid themselves back for the artist's small portion of the overall revenue. While recording artists create the value of the recorded music commodity, they are not necessarily compensated for the work that they perform but rather are treated as independent contractors that lose all autonomy through their contractual agreements.

The degree of creativity supposedly separates the studio musician from the recording artist. However, the difference between the degree of creativity of the recording artist and that of the studio musician is often negligible, if it exists at all. First, if everyone could perform the task of the studio musician, session gigs would provide lower wages for musicians because there would be an increase in supply of capable musicians. Second, the studio musician is most certainly adding value to the recorded product in the same way as the recording artist, if on a different scale. Studio musicians use the means of production (i.e., their instruments, the studio, and the written score) to add value to the final product. The contractual relationship between the recording artist and the label does not stipulate that the recording artist must perform any more or less labor in the studio than the session musician.

This leads to my final point that negates the creative difference between recording artists and studio musicians: composition and improvisation. While the final recorded product is under the name of the recording artist, this does not mean that that artist is the author of the composition. In fact, popular music stars rarely have the autonomy to play the music that they write. Rather, popular music stars usually play the music that a professional songwriter or producer composed. This means that in the studio there is little difference in the creative role of the recording artist and the hired studio musicians. Additionally, the producer and A&R staff often dictate the way that both the recording artist and studio musicians perform a particular song. For a recording artist performing the composition of a separate songwriter, the only thing that makes it their music is that their name is prominently displayed on the song.

Recording artists are the commodity, with their names, images, and stories emblazoned on the medium, while session musicians "are not prominently highlighted on the album cover, packaging, or advertising" (Krasilovsky et al. 2007, 17). By holding the labor of the recording artist separate from the labor of others within the recording industry, record labels develop a fundamental distinction that is used to mystify and mythologize the creative energy of recording artists' labor, even though their labor is rarely qualitatively different from that of other musicians. Recording artists are bound to the success of the recorded product, and they must promote the sale of their music, whereas studio musicians' relationship to the success of the album, for the most part, stops at the door to the studio.

For many musicians, there is an ideology that makes becoming recording artists (i.e., "getting signed") the end goal, showing that one has "made it" in the music business. "A lot of kids dream of becoming recording artists. . . . But it's often a long, tough road between the dream and the recording contract" (Krasilovsky et al. 2007, 14). Without fully realizing the material consequences of signing a record contract, musicians want to actualize their aspiration to be recording artists. Many musicians guide their careers from a young age toward what they think will result in them getting a record contract. Music business books help to mythologize the record contract by offering advice on the procedures to follow in order to obtain one.

One of the most valuable lessons that any band or musician that has dreams of being signed can learn from these books is how to set up their band as a business;[4] as a result, many bands incorporate in order to get the full legal protections of "corporate personhood." As small businesses, these bands try to develop their "brand" and business to a point where they seem attractive investments for a record label. They do this by developing a following, selling albums and merchandise, and having an active gig schedule.

Bands hope to be "discovered" by an A&R representative (Krasilovsky et al. 2007) while playing shows—sometimes this is fostered by playing at show-cases or through inviting A&R staff to a show. When a record label deter-mines that signing a band is an attractive proposition for one reason or another, the label enters into a contract with the band's business manager. However, it is important to note that these contracts do not act as business takeovers but rather as contracts between separate entities—the band and the label continue to exist separately (Krasilovsky et al. 2007); this creates a relationship where the signed act is a subcontractor to the label. Under this contractual relationship, labels owe nothing beyond the initial advance to their artists in return for their labor.

The music industry pressures bands to become a business from the start. This involves everything from hiring a manager and incorporating the band to signing contracts with venues and selling merchandise. Since record labels want to contract with independent business entities, bands are more than happy to develop their structure in that direction. However, one thing that bands rarely do is take out small-business loans to get the band established. Arguably, record labels act like lenders when they give their artists advances. Musicians could ostensibly launch their careers using loans as easily as sign-ing a record contract. It is for this reason that I think both the ideology of being signed and the contract itself create particular relations of production that allow the major record labels to extract surplus value out of recording artists.

POLITICAL ECONOMY OF MUSICIANS' LABOR

While music has continually undergone transformations in its media forms, the structural relations of musicians' labor have remained relatively un-changed. Apart from the role of session musicians, most musicians make and perform music in a relation of production that is technically similar to that of a petite bourgeoisie artisan. However, there are deep contradictions in the concept of intellectual property ownership that undercut the ideas of author-ship and artistic autonomy. In other words, recording contracts designate recording artists as petite bourgeoisie, but the terms of the agreement struc-ture recording artists as workers.

The political economy of music has gone through some massive changes as music has shifted from a ritualistic cultural act to a commodified cultural good. Jacques Attali (1985) distinguishes four distribution networks that cor-respond to four structures of economic production; these networks of distri-bution act as waves[5] as the remnants of previous networks coexist with contemporary networks. First, "sacrificial ritual" is the network of music that Attali associates with symbolic societies—these are societies that Attali says use rituals of symbolic sacrifice to keep outside forces from ripping them apart. This network of music production is found in precapitalist societies, which Attali associates with feudalism. During the period of "sacrifice," music does not have a place within the economy, it is something experienced and enjoyed by people at festivals and on the streets. Second, "representation" involves the first injection of capital into music as people begin to pay to see music performed. The economic mode of production for representation was early capitalism; initially, nobles paid musicians to perform. This network of distribution later adapted as the growing bourgeoisie used their money to "represent" the power of the nobles by paying admission to orchestral concerts. "Representation" was dominant from the end of Feudalism until the late nineteenth century. The stream of revenue from live performance for musicians comes from "representation."

Third, "repetition" is a result of recorded music, which allows for a "new organizational network for the economy of music" (Attali 1985, 32). Recorded music spatially and temporally disentangles the musician from the music, but the structure of record contracts maintains the material link between artist and commodity. The recorded commodity is symptomatic of industrial capitalism and mass media and has been dominant from the early twentieth century until today. Finally, Attali projects a fourth network of "composition" that "would be performed for the musician's own enjoyment, as self-communication, with no other goal than his own pleasure" (1985, 32). A network of "composition" might be brought about by digital technology, but such a potential outcome is beyond the scope of this project.

Although music distribution networks have changed with structures of production, the structural relationship between artists (labor) and capital (record labels) has been slow to change.

In order for musicians to satisfy their basic needs by playing music, society must have a social division of labor. Musicians must be able to exchange their music for other things, mainly to meet their needs of subsistence. The act of performing music becomes a commodity for music listeners as a result of "useful labor, i.e. productive activity of a definite kind, carried on with a definite aim" (Marx 1992, 133). This useful labor occurs in a capitalist system of exchange because whether music is recorded, written, or performed live, the music listener pays the musician to listen to the music. As long as this is a direct relationship, the price that the listener pays to the musician equals the value of the labor that the musician needs to reproduce another day of labor. However, the musician is not doing this in a manner that accumulates capital because the value of the music includes only the necessary labor[6] for the reproduction of the worker. Since the musician easily owns his or her means of production (namely, an instrument), there is no room for the involvement of capital to initiate the process of the accumulation of capital during a kind of production that relies strictly on live performance. In this way, the musician's labor is in a simple exchange relationship[7] because he or she exchanges labor for money only in order to buy commodities in another form.

With the advent of phonographs and gramophones, recorded music became possible; this made studio performances available for consumption at remote locations. Recorded music structurally altered the means of production because it changed the relationship between performer and audience. Owning one's instrument no longer constituted the extent of the means of production under a distribution network based on "repetition" (Attali 1985). In order to record music, musicians have to use expensive recording studios, print the music onto a medium, and distribute the recorded commodity over large areas. Additionally, this kind of production separates the worker through time and space from the consumer; therefore, "repetition" creates alienation because musicians no longer directly interact with their fans. Expensive means of production also enable capital to play a role in the production and reproduction of music. In return for the ability to record and distribute music, musicians enter into contracts with record labels.

Record contracts enable the creation of surplus value in recorded music[8] because they stipulate that musicians produce music in exchange for access to the means of production; these contracts also change a musician's labor position from petite bourgeoisie to that of a worker. Marx's labor theory of value helps explain the way that musicians create surplus value by expending labor power in excess of their necessary labor (Marx 1992, 325); for Marx "surplus-value originates from the difference between what labor gets for its labor-power as a commodity and what the laborer produces in a labor process under the command of capital" (Harvey 2010, 125). Musicians sell their labor power for a price, but the record label uses that labor power to create more value than it has paid its workers. The musician, "like the seller of any other commodity, realizes its exchange-value, and alienates its use-value" (Marx 1992, 301). The musician's use value for the record label is the total amount of value that the musician makes in a day; surplus value is the value derived from the musician's use value minus the exchange value paid for his or her labor power. In this way, drawing on Marxian economic theory, profit in the recording industry is the value that the record labels possess after paying the musicians (variable capital) and the cost of the means of production (constant capital).

This is precisely the way that labor operates for session musicians because they receive a wage based on performing a specific amount of music. Session musicians are hired for a price that they think will cover their cost of reproduction. In wage terms, the rate can seem quite high (according to the American Federation of Musicians, the union rate in Los Angeles is $380 for a three-hour session), but sessions may be rare for a musician. Therefore, session musicians receive this wage under the knowledge that they may only have two sessions a week. Record labels have a good idea that their profits from session musicians' labor will exceed whatever they pay in union rates.

However, in the case of a recording artist, the musician never receives payment for his or her labor, so the means of production under recorded music does not create a system of wage labor in the recording industry. Instead of recording in exchange for a wage, recording artists agree to perform certain duties and relinquish certain intellectual property rights in exchange for a cash advance and their label's technical expertise in recording and distributing their music. Musicians signed to a recording contract are still considered contractually independent business entities. When recording artists receive their advance to record an album, they have to take some money out to pay themselves for their necessary labor while they record and promote their music. In this way, recording artists do accept a form of compensation for their daily needs; however, they are expected to pay back the advance from the royalties that they earn from selling their music. In other words, recording artists pay themselves the cost to reproduce labor with capital that they do not have. Depending on how well an album does, the record label decides whether it would like to record another album for the recording artist (Hull, Hutchison, and Strasser 2011; Slichter 2004). If the label decides *not* to record another album, the recording artist is typically not permitted to record with a different label unless both entities mutually agree to break the contract. While technically remaining independent from the record label, recording artists lose all autonomy as an independent entity outside of the record label because of a record contract.

The contractual obligation of record contracts designates the material relationship between a record label and a recording artist: capital and labor.[9] However, instead of selling one's labor as a commodity, recording artists exchange their capacity to produce music for the technical equivalent of a loan. Labels argue that this relationship maintains the autonomy of the musician in the creative process; however, A&R staff and record producers play an important role in the creative process that transforms a recording artist's music to fit the needs and vision of the label. In actuality, signing a record contract eliminates all autonomy for the contracting musicians. What at first glance appears as a mutually beneficial process for both musician and label is a parasitic relationship where the record label takes the copyrights of its artists (i.e., the artist's future profits) in exchange for the money to record an album. While many musicians may gain some enjoyment out of this relationship that is unquantifiable in terms of an exchange value (e.g., the feeling of being a rock star or being flown around the country to promote an album), the fact remains that the value of their property rights and their labor power is being expropriated by their record label.

At this point, some may argue that the contractual relationship of recording artists negates this application of Marx's labor theory of value, but I contend that the contractual relationship is ideological. While the law stipulates that recording artists are independent contractors, musicians are the labor of the recording industry. Contracts are defined by law, and law is ideological (Gramsci 1971). Record labels rely on the ideology of the artist specifically to perpetuate the notion that a recording artist's labor is somehow different from all other labor; this is just not the case. Record contracts act to conceal the real relations of production by establishing the efficacy of copyrights over wages.

While I am arguing that the structural relationship of labor in the music industry has not changed through technological mediations, I am not arguing that this labor has been unchanged. In fact, musicians' labor has continually changed over the past century. Part of the evolution of labor is evident in labor conflicts that occurred within the American Federation of Musicians (AFM), particularly over the usage of recorded music in movie soundtracks. As Jon Burlingame describes in *For the Record* (1997), the rise of recorded music ultimately created a conflict between studio musicians in Hollywood and live performance musicians across the country as Hollywood movie and television producers began using recorded music on their films and shows. At issue in this conflict was a fundamental shift from a system where live performance was ubiquitous as live bands performed movie scores to one where machinery replaced the labor of live performers. Recorded film scores eliminated the need for live performance labor and destroyed the jobs and job prospects of musicians across the United States. While Burlingame trivializes the tactics of AFM's president in making recorded music prohibitively expensive, the reality of the moment (1940s–1950s) was that the majority of musicians felt that Hollywood studio musicians were putting them out of work. While Hollywood studio musicians were among the few musicians actually making a living playing music, this minority of musicians was using technology to put others out of a job. These types of conflicts continue today as disc jockeys (DJs) have replaced cover bands over the past four decades and digital audio workstations such as Pro Tools, Garage Band, and Reason have replaced the need for studio musicians. However, these shifts always change the labor of session musicians and do not really change the position of recording artists.

LIVE PERFORMANCE AND OTHER STREAMS OF REVENUE

Of course, few musicians depend entirely on making money from recorded music. The recording industry is only one part of the broader music industry, and musicians work in different ways to generate revenue and income. The primary source of income for most musicians comes from performing music live (Hull, Hutchison, and Strasser 2011, 143); this has remained unchanged since the stage that Attali describes as "representation"—the end of feudalism. Since live performance is the moment when musicians have contact with their audiences, it is also an opportunity for them to sell merchandise. Through live performance, musicians have a considerable amount of control over their labor because they use their management staff to negotiate contracts with venues, promoters, and ticket agents to stipulate the terms under which they will perform. It is in this sense that musicians resemble petite bourgeoisie regardless of whether or not they are signed to a record label because they produce goods that they sell to consumers with little mediation (i.e., alienation).

When bands perform live,[10] they use their management teams to create a contract that meets their desired price for their labor. Generally, fans pay admission to see a band perform; however, it is up to the band and the venue to decide what will pay for the band's performance. The contracts that bands agree to in order to perform vary depending on the size of venue and the popularity of the band. Playing at a small bar may mean that a band gets the money made (or percentage of that amount) at the door (i.e., a percentage of the total revenue from admission). Different arrangements can be made if the band is likely to bring in a significant crowd. In those cases, the band may receive a guarantee plus a percentage of the door (i.e., revenue from tickets sold). Venues that are also bars make more money from alcohol and food sales by having a band perform, so they are sometimes willing to share some of that revenue (in less lucrative deals, bands must pay a percentage of the door to the bar). At other times, the performers might just get drinks and dinner paid for in return for the performance. Generally, the band will also have to pay for sound and any other support staff (e.g., lighting) from the amount agreed upon, or that support staff could also be paid a percentage from the door. These contracts are highly negotiable and can stipulate everything down to how the venue is to interact with the musicians in the band.

While musicians own part of the means of production in live performance (i.e., their instruments and any sound equipment), they rarely own the venues at which they perform, so the venue owners take the position of capital in live performance. However, the relationship between a band and a venue is often mutually beneficial in a way that complicates the application of Marx's labor theory of value in this situation because the labor theory of value applies to wage labor and this is a contract between two firms who then split up the risks and rewards. The labor theory of value would kick in if the venues had a "house band" or the band was paying session musicians for their performances. The difference between what a band is paid for a performance and the amount that the venue makes from a concert is the profit, but the profit is usually generated from the sale of alcohol, food, or service fees that are not directly derived from the band's labor. While bands are unaware of the amount of money that moves through the venue that night, they have a general idea of how much revenue ticket sales generated from their show. Since venues depend on selling alcohol and food to generate their revenue, if a bar does not have a musical performance, there will be less revenue earned. Live performances create two different yet related primary revenue streams: from an audience through tickets and from food/alcohol.

Since live performance is when bands have the most contact with their fan base, it is a point when bands sell a considerable amount of merchandise; it is also an opportunity for bands to fully embrace their position as petite bourgeoisie by selling goods created by the labor of others. Fans buy everything from T-shirts to key chains, posters, and albums while at the concert, but bands rarely make their own merchandise; rather, they rely on global commodity chains to produce their merchandise at a cheap cost. When a tour manager plans a tour, the known revenue paid to a band for the performances pays for the costs of touring (hotel arrangements, paying for the road crew, transportation, food, etc.); the profit from the sale of merchandise is profit for the band. These alternative sources of income for musicians are important for the discussion at the end of this chapter in the way that "360" deals have changed revenue for recording artists.

FROM DISINTERMEDIATION TO DE-ALIENATION?

While the above descriptions illustrate the way that bands have operated as labor within the music industry over the past sixty or so years, the digital transformation created a definite shift in the way this system works. From increasing contact between bands and fans to allowing musicians to access better information on contracts, the Internet changed the way labor functions in the music industry. Most important among these changes is the ability of musicians themselves to promote their own music through websites and social media. Disintermediation—the elimination of intermediaries from distribution—created the *potential* for musicians to have better connections to fans, but the results of these greater connections have not changed the material relationship between musicians and their fans.

Bands have always had a problem connecting with their fans. When first starting out, a band used to acquire exposure only by playing in front of audiences and trying to get people to shows through local marketing (e.g., distributing flyers). After playing as an opening act several times, bands begin to develop a new fan base. Local record stores carry the music of local bands along with some merchandise, so when people walk into the stores they can find local bands' music. These stores also carry music magazines. Most of these magazines are genre specific and often include articles that focus on independent music. Fans of a certain genre can keep up with that scene by reading these publications. There are popular magazines with large circulations that cover multiple genres, such as *Rolling Stone* and *Spin*, and genre-specific ones such as *NME*, *Down Beat*, and *XXL*. These magazines also alert fans to new albums by already established acts. Another important aspect of the local record store is the clerk, who generally has a good idea of what is going on in a given music scene (partially from reading the magazines sold at the store). Before the Internet, magazines, newspapers, and promotional items put out by venues (e.g., show lists and billboard signs) were generally the only way to keep informed about what bands were playing in a local area. Labels and bands used street teams—"local groups of people who use networking on behalf of the artist in order to reach the artist's target market" (Hull, Hutchison, and Strasser 2011, 271)—to help promote shows. For smaller venues and bands, promotion for shows often happened through flyers at the venues, supermarkets, music stores, and the like; for larger acts, there might be radio ads and newspaper ads. This system limited music to small circles of fans and subcultures that were familiar with particular scenes; it did not allow bands to easily cross over into wider audiences.

In the mid-1990s, the Internet changed band–fan communication as bands began sending newsletters about upcoming shows, albums, and general information to their fans via email. At shows, bands began passing around sign-up sheets for their listservs as a way to stay in contact with fans; since fans voluntarily signed up for band listservs, bands knew that they were contacting people who were at least partially interested in their performances and news (Harding 2010; Lieb 1994). The Internet enabled fans to keep up with their favorite bands as easily as opening an email message.

Internet websites dramatically changed the way that bands communicated with fans in the late 1990s. A band website can have news, concert information, biographies, promotional items, and information about upcoming albums. The Internet not only allows bands to publicize, promote, and interface with their fans, it also allows the direct publication of their music, sometimes free, sometimes with various restrictions. Bands can also place streaming music on their websites to give site visitors a sample, and some bands have made their music available for download from their sites (for free or for a charge). There is usually a "store" link on band websites where visitors can purchase merchandise and music. One recent trend is for bands to make exclusive performances available on their websites for a fee. Another avenue that some bands have begun to explore is live streaming of concerts or imbedded YouTube videos on their websites. All of this has made it easier for bands and musicians to promote and publish their music and contact their fans.

While websites and mailing lists made it easier for bands to stay in contact with their fans, social media had the largest impact on band–fan relations as social networks altered the ways fans can choose to interact with their favorite bands and vice versa. Social networks "have consequences for how people get to music, and for how music gets to people" (Jones 2002). According to the Artist Revenue Streams project by the Future of Music Coalition, 78 percent of musicians surveyed acknowledge that emerging technologies have allowed them to communicate directly with fans (Thompson 2012). While 43 percent of Artist Revenue Streams participants have a blog or website, roughly 69 percent use the Internet to promote their music and connect with fans (Thompson 2012). Beginning in the mid-2000s, MySpace immediately became a useful tool for bands as it let them communicate directly with fans, post streaming music, and post a calendar for free. With a MySpace page, bands no longer had to spend money and time developing a website or paying someone else to do it because MySpace allowed them to do this for free. Facebook is similar to MySpace and has increased in popularity recently, with 44 percent of musicians claiming to use it to promote music (Thompson 2012). Both MySpace and Facebook allow bands/musicians to create events and invite friends. The functionality of Facebook allows bands to target event invitations directly to specific locations or populations. Both social networks allow bands and musicians to have free websites. Bands have incorporated social media, email lists, and websites as part of their marketing toolkit, but have these new marketing strategies changed the ability of independent artists to compete with major record labels?

With all the potential that social networking provides musicians for circumventing the major record labels, there are two problems with the use of social networking sites as a means of distributing music to fans. First, social networking sites (e.g., Facebook, MySpace, and Twitter) do not allow users to upload music that, in turn, can be downloaded free by fans. In order to use these networks for digital distribution, musicians have to either use a pay system internal to Facebook or MySpace or create a website they can link to on the social networking site. Second, before musicians can upload their music to social networking sites, they must first establish that they are the original authors. These mechanisms create barriers to independent musicians for getting their music on the Internet alongside music by musicians signed to major record labels.

As these websites became places for musicians to interact with their fans, some scholars have argued that this is also creating a more "participatory culture" (see Jenkins 2006). However, this scholarship is shortsighted about the material relationship that is occurring in such environments. "Our artists," former Sony executive Fred Ehrlich argues, "found online promotion to be an exciting tool which empowers music fans to communicate directly with their favorite artists" (Hatch 2000). These websites hardly "empower" music fans, but they do provide new opportunities for musicians to contact and inform their fans. Mark Andrejevic establishes, rather, that this idea of interactivity only helps capital to solidify its position through surveillance. "Somewhere in the mix, the positive associations of interactivity as a form of two-way, symmetrical, and relatively transparent communication (in the sense of knowing where the information we send is going)," Andrejevic explains, "have been assimilated to forms of interaction that amount to little more than strategies for monitoring and surveillance" (Andrejevic 2007, 5). Companies use data collected through monitoring and surveillance to more accurately market music and other products to their consumers. Online networking and surveillance not only helps independent bands reach audiences, but it also empowers record labels to hone their marketing acumen. On the other hand, from the perspective of a musician, the Internet created more opportunities for keeping fans informed and boosting attendance at shows; this resulted in greater autonomy from recorded music and record labels because the importance of the recorded commodity has declined.

As a result of the increased connectivity between musicians and their fans, many commentators thought that musicians could begin to circumvent record labels altogether (see Burkart 2010; Burkart and McCourt 2006; Knopper 2011). For instance, Patrick Burkart contends that musicians and "fans are challenging the current business model of digital distribution" (Burkart 2010). The Future of Music Coalition also encourages new markets for independent music online. However, the fact remains that major record labels and their subsidiaries have maintained the position of gatekeepers in the music industry throughout the digital transformation. Disintermediation created the potential for musicians to distribute music directly, but internal limitations to their interactions on the Internet maintain power imbalances in the larger music industry.

Chapter Six

Victims, Musicians, and Metallica

There was a twofold victimization of musicians by the recording industry during the transition to digital music. First, the Recording Industry Association of America (RIAA) and the major record labels contended that file sharers victimize recording artists because file sharing cuts into their artists' means to make a living. This victimization created the ideology of an industry in crisis, and this was used by the RIAA to lobby Congress to change laws governing intellectual property and stop consumers from downloading free music. This is an ideological or even fictional way of victimizing the musician. The real victimization occurs when the major record labels use new laws and their claimed loss of profits to further extract surplus value from their recording artists' labor, mainly through the implementation of so-called 360 deals.[1]

VICTIMS OF FILE SHARERS

When Lars Ulrich, drummer for Metallica, testified at the Senate Judiciary Committee hearing entitled "Music on the Internet: Is There an Upside to Downloading?" (Hatch 2000), it was not to argue that the Internet had somehow liberated his band to connect with fans, but rather that Metallica's own fans were stealing from the band. It turns out that as the fourth all-time highest-selling band with the number one selling album in the SoundScan era (Nielsen 2011), Metallica desperately needed to find ways to sell more music. Ulrich's testimony accentuated the RIAA's argument that the major labels' recording artists were being unjustifiably victimized by teenagers downloading music on peer-to-peer (P2P) file-sharing programs (at the time, Napster and Gnutella). In order to construct the public fervor necessary to legislate against the use of P2P programs, the RIAA and the major record labels made it a point to demonstrate that their recording artists were victimized. To do this, the RIAA did not use some abstract musician on their labels. Rather, they enlisted a rockstar that downloaders could recognize—Lars Ulrich. By publicly arguing that downloading music for free was creating hardships for the very recording artists that file sharers venerated, the RIAA hoped to change attitudes about file sharing and the rules that govern copyright.

In order to get the law changed, the RIAA portrayed its best-selling artists as victims of file sharers. According to Jonathan Simon in *Governing through Crime* (2007), one of the most effective tactics for passing legislation is to construct a victim that must be protected. "Lawmaking rationalities," Simon states, are best articulated by "identifying broad sectors of the American population through subject positions that help elaborate the purpose of legislation . . . including which 'enemies' must be confronted by the government to protect citizens" (2007, 79). While Simon's focus is on victims of violent crime, the construction of a victim of copyright violations follows the same pattern. Simon's three corollaries to the idea of "governing through crime" are:

1. Crime has now become a significant strategic[2] issue.
2. We can expect people to deploy the category of crime to legitimate interventions that have other motivations.
3. The technologies, discourses, and metaphors of crime and criminal justice have become more visible features of all kinds of institutions, where they can easily gravitate into new opportunities for governance. (Simon 2007, 4–5)

Together these three claims point to a system in which legislation, judicial rulings, and policy are aimed solely at protecting victims from crime. Before the law can intervene to protect victims, people must construct victims. Therefore, it is not enough for copyright owners to decry file sharing in the spirit of fairness or tradition. Copyright owners must demonstrate that the act of sharing files of copyrighted material is no different from stealing. Free music available via file sharing is not only the consequence of a change in medium, but to the major record labels, it is also a crime; however, as I demonstrated in Part II, file sharing is not necessarily a crime.

Major record labels made this argument about property theft with the goal (1) to create new laws and (2) to implicate the public as criminals to deter people from file sharing. Even though copyright law clearly prohibits the unauthorized *commercial* reproduction of copyrighted material (among other activities, such as recording a song without the composer's permission), the major record labels and the RIAA have repeatedly referred to file sharers as pirates. Furthermore, there is a clear legal line that separates copyright violations and theft (*Dowling v. United States*, 1985), but the recording industry is unconcerned about this legal nuance in its descriptions of file sharing as theft. By constructing the piracy panic narrative, the RIAA was positioning the issue of file sharing as a crime, and politicians "turn to crime as a vehicle for constructing a new political order" (Simon 2007, 25). The important point here is that the recording industry controls the discourse in a way that stigmatizes file sharers as criminals and signifies recording artists as their victims.

At the Senate hearing, every speaker and senator, excluding Roger McGuinn (singer and guitarist for the Byrds) and Hank Barry (CEO of Napster),[3] premised their position on the axiom that musicians (in their words, artists) must be compensated for their work through intellectual property. This premise is built on two assumptions. First, they assume that artistic labor must produce a good that can be sold directly by the artist, rather than the worker being compensated for hours put into labor. They were not making the point that recording artists should be compensated for their labor power the same way that an employee of one of the labels is compensated. In fact, during the hearing, Fred Ehrlich, president of new technology and business development at Sony Music Entertainment, explained at length how many jobs record labels produce for every album that they release. He then went on to explain that the few top-selling recording artists pay for all of the jobs at record labels that go into producing unsuccessful albums because most albums never have their cost recouped, and the few top-selling albums more than make up the difference. However, Ehrlich does not explain why unprofitable artists do not receive payment in the same way as the people that clean their studios.

Second, they assume that musicians would only produce music under a system that pays them for the music that they produce. This assumption points to an understanding of music that is built upon the networks of music that Attali (1985) describes as "representation" and "repetition" because it assumes a capitalist end. Under a capitalist economy, they contend, there is no incentive for musicians to create music without a mechanism for them to commodify and monetize their music. What this fails to see is the innumerable people who write and perform music with no intention of selling it; these could be people who perform music for ritual or social purposes (e.g., sitting around a campfire playing a guitar) or for their own enjoyment. Yochai Benkler suggests, in *The Wealth of Networks* (2006), that there are numerous reasons that people produce content outside of a market system. It is important to keep in mind that regardless of what the minority of rockstars earn, the vast majority of musicians actually make little to nothing from their musical work.

While Article I, Section 8, Clause 8, of the Constitution (the "copyright clause") provides for the protection of intellectual property, it is interesting that Sen. Patrick Leahy mentions this article in his opening remarks rather than any other section of the Constitution. By beginning with the copyright clause, Leahy is already making a judgment about the role of intellectual property without considering the potential benefits of making music available free online. This position gives the copyright hawks legal ground during the hearing to assess whether the intention of protecting copyrighted material is possible instead of questioning whether the copyright clause has anything to do with the matter at hand. I bring this up because, as I demonstrated in Part II, copyright policy is complex and always evolving, and when copyright does change, it is changed through multiparty negotiations that include only those groups with a stake already written into the copyright law. Sen. Leahy's insertion of copyright law into the debate set the stage for the recording industry's claim that P2P technology allows music fans to victimize recording artists.

At the center of this discussion was the view that Napster was not only allowing fans to get music without compensating their favorite recording artists, but it was doing this without the consent of the artist. Ulrich was the main champion of this position, contending that Metallica was the unknowing victim of people carelessly stealing their music. As Ulrich explains:

> I do not have a problem with any artist voluntarily distributing his or her songs through any means that artist so chooses. But just like a carpenter who crafts a table gets to decide whether he wants to keep it, sell it, or give it away, shouldn't we have the same options? We should decide what happens to our music, not a company with no rights to our recordings, which has never invested a penny in our music, or had anything to do with its creation. A choice has been taken away from us.
>
> With Napster, every song by every artist is available for download at no cost, and of course with no payment to the artist, the songwriter, or the copyright holder. If you are not fortunate enough to own a computer, there is only one way to assemble a music collection the equivalent of a Napster user— theft. Walk into a record store, grab what you want, and walk out. The difference is that the familiar phrase "files done" is now replaced by another familiar phrase, "you are under arrest." (Hatch 2000)

In this passage, Ulrich moves from a cool indifference to the decision by other artists to release their music free online to a comparison of using Napster to stealing from a record store. The transition between the two positions does not account for Napster users who download music from musicians who freely share their music. Starting his argument by drawing comparison to a carpenter echoes parts of chapter 1 of *Capital, Volume 1* (Marx 1992) when Marx discusses the fetishization of building a table (i.e., a carpenter must build the table), but Ulrich quickly makes it clear that his argument is the antithesis of Marx's critique. In contending that Metallica "should decide what happens to our music, not a company with no rights to our recordings," Ulrich is claiming that Napster itself is victimizing Metallica. However, few recording artists have the same control over their work as a carpenter because companies (i.e., major record labels) control what these musicians can do with their music. Metallica actually has more control or rights over their music because of their ability to renegotiate their contract with Elektra, a subsidiary of Warner Music Group.[4] On the other hand, this is the entire point; Ulrich claims that Metallica are the victims of fans literally stealing their music—no different from stealing music from a record store—without any concern as to the actual control that recording artists have over their music.

Metallica's position in the recording industry as one of the top-selling acts of all time that used their position to negotiate more rights from their label makes them the ultimate poster child for the recording industry to demonstrate the victimization of recording artists by P2P users. On April 14, 2000, Metallica became the first recording artist to file a lawsuit against Napster for allegedly encouraging "users to trade songs and recordings without the band's permission" (Segal 2000). The difference between Metallica and most recording artists is that they "own their masters and control their catalog" (Thigpen and Eliscu 2000); since Metallica had a lot more to lose than other recording artists, having them at the center of the battle over P2P based on their own interests gave the recording industry a popular band through which to voice its concerns. It is also important to note that few other musicians were openly voicing concerns about Napster; this is at least partly because of the structural position of Metallica in relation to the majority of other recording artists. However, regardless of Metallica's position as one of the top-selling bands of all time, they wanted to let people know that downloading their music free still hurts them. *Rolling Stone* cites Metallica's attorney Howard King as saying, "It's not just a bunch of corporate thieves worried about their money. Metallica want people to know that they are being hurt. Piracy affects artists" (Thigpen and Eliscu 2000).

To drive his point home about how Napster and its users victimize Metallica, Ulrich provided data on the sharing of Metallica's music online to the Senate Judiciary Committee. "In fact, in a 48-hour period where we monitored Napster," Ulrich claims, "over 300,000 users made 1.4 million free downloads of Metallica's music" (Hatch 2000). In a post-iTunes world, 1.4 million downloads would equate to $1.386 million (at 99 cents/song) in revenue over a forty-eight-hour period—note that a new album was not released around this forty-eight-hour time period that would drive this degree of consumption. Ulrich declared that Metallica should be compensated or have the authority to prohibit their music from being downloaded. However, he fails to reconcile this logic with physical album sales. As I mentioned above, Metallica is the fourth highest-selling musical act since 1991 and has the number one selling album during that period. Contending that the band could possibly see regular sales of 1.4 million units every two days seems disproportionate, even for one of the highest-selling recording artists. These data make it difficult to entertain the idea that a download is equivalent to stealing a physical album.

Yet it was not Lars Ulrich's pleading to the Senate Judiciary Committee that changed the debate about Napster's legality, but rather the news media's acquiescence and acceptance of the argument that downloading music was in fact piracy and equivalent to stealing from a fan's favorite artists. Following a similar line of reasoning to many of the speakers and senators at the Senate hearing, the media appropriated the idea that fans were causing an actual concrete harm to their favorite bands. Far from reporting the facts about P2P file sharing, reporters were quick to take a position on the topic that favored Ulrich's and the RIAA's position. The following is an excerpt from a *Washington Post* article by Jonathan Yardley.

> Metallica is right, of course. Musicians have as much right to earn a living off their work as do all others in all other lines of work. . . .
> The basic issue is whether artists have a right to the fruits of their labors. The answer, under U.S. copyright law, is emphatically in the affirmative. But in the ever-changing world of the Internet, assuring that right is even more difficult than it was in the past. Bringing Napster to heel actually is likely to be fairly easy, because, as a report in the Wall Street Journal points out, "much of the [music-distribution] system rests on Napster's computers, and can therefore be easily shut down," a process that in fact seems to have begun in response to Metallica's lawsuit. But a newer and even hipper music-copying program, Gnutella, "relies on no single central machine," allowing its users to "connect to each other directly in a constantly mutating network." In other words, Metallica—acting as point man for all artists and writers and others of its ilk—may win its battle against Napster, but the war has only just begun. (Yardley 2000)

While Yardley's claim that musicians "have as much right to earn a living off their work as do all others" is correct, his "emphatic" assertion that Napster is about whether "artists have a right to the fruits of their labors" deserves careful interrogation. This argument assumes that P2P programs are useful only to the extent that they circumvent copyright law. Yardley's contention is that since P2P is useful only to circumvent copyright law, P2P programs interfere directly with the ability of artists to earn a living. News reports at the time were becoming heavily entrenched in the position that file sharing was equivalent to stealing (see Kedrosky 2000; Thigpen and Eliscu 2000; Yardley 2000). The news media went from ogling over the new technology that allows people to obtain music freely to becoming the defenders of musicians against those fans that would put their favorite musicians on the street in order to listen to some music free.

Most of the claims that marauding college music fans victimize musicians were driven by the RIAA. In fact, while not an official member of the Senate Judiciary Committee hearing panel, Hilary Rosen, then CEO of the RIAA, seems to have orchestrated the entire hearing: not only was the panel of witnesses heavily organized in favor of the RIAA's position, but during questioning Sen. Hatch actually called Rosen forward to "come up to the table, as well, so she can answer questions if anybody has them" (Hatch 2000). Why have Rosen come to the table to answer questions, but not a well-known legal scholar such as Lawrence Lessig? By requesting that Rosen be part of the formal hearing, Sen. Hatch demonstrates that the RIAA's perspective was of critical importance to the government establishment and that he wanted to make sure that all questions were answered to its satisfaction. As part of the RIAA's campaign, Rosen made appearances on television and wrote op-eds in newspapers. In one such op-ed, she flatly stated, "Technology may make stealing easier. But it doesn't make it right" (2000). She continued that "When creative products are illegally duplicated and distributed without the consent of or compensation to the artists who produce them, their ideas are devalued and their voices are silenced" (Rosen 2000). Rosen contends that the technology of P2P is good only for stealing copyrighted work and that this is directly harming the ability of artists to receive an income. However, the RIAA represents record labels, not artists, and Rosen was never fully held accountable for this position since she could always point concretely to Metallica—despite the fact that few other recording artists were willing to criticize Napster and many supported it. Simply writing off file sharing as lost income to recording artists fails to address the material interests of those very same musicians.

If, on the other hand, we take seriously Yardley's contention that musicians "have as much right to earn a living off their work as do all others" (2000), then we can come to an entirely different set of conclusions than the protection of copyrights because most contracted musicians do not own their copyrights. Yardley maintains that musicians are workers and that they deserve to be compensated for their work. When he makes a leap to connect copyright to the interests of those musicians, Yardley makes a glaring misstep in logic. Recording artists are workers who deserve to be compensated for their labor, but the reason that they are not being compensated for their labor is the insistence on the part of record labels that they be paid royalties on the sale of their music instead of as workers who sell their labor power for a wage. The way that the recording industry is structured causes recording artists to lose income. Rather than victims of piracy, musicians are the victims of capitalism. Recording artists do not have a lot to lose from fans file sharing because for most artists, the royalties that they would earn would only go toward paying back their advance; in fact, it is the major record labels that have a lot to lose as musicians find new ways to sell their labor power to their fans. The idea that fans victimize recording artists when they share music allows record labels to re-create record contracts that further enchain recording artists to their labels.

360 DEGREES OF SURPLUS VALUE

The Internet provides musicians with new ways to address their fans; this has the potential to change the entire political economy of the music industry as musicians can use the Internet to direct their fans to live performances and merchandise. If recorded music becomes devalued and operates more as a promotional tool than as a commodity, then the recording industry would lose nearly its entire business; but as Part I demonstrates, this has not happened. To compensate for the potential change in which commodities get valued, the recording industry has used the rhetoric of lost income due to piracy to renegotiate the ways it profits from artists. This is one of the ways that the recording industry deploys crime to initiate change in the relations of production. Record labels point to the way that fans are victimizing their favorite musicians to exploit recording artists' labor further by appropriating new aspects of it. The argument that recording artists are the victims of illegal file sharing helps to construct and support the major record labels' position that the recording industry is in crisis. This becomes an ideological position as it acts to obscure the fundamental relationship between capital and labor. As the general public and the political apparatus accept this ideological position, the piracy panic narrative creates political capital for the major record labels that they use to reconstitute their relationship with their recording artists through the establishment of so-called 360 deals.

Because of the supposed decline in music revenues and profits, record labels are seeking ways of exploiting more of their recording artists' labor beyond expropriating artists' copyrights. Major record labels recognize that there has been a reversal in their artists' income: the largest proportion used to be from record sales, but now the largest source is "concert performances and merchandising" (Forest 2008, 167). The main tool that record labels developed to secure additional revenue, besides suing their consumers, is the "360 deal." A 360 deal is a "deal between a label and artist that involves the label receiving revenues from more than just the recording aspects of an artist's career, but also probably live appearances, music publishing, and merchandising" (Hull, Hutchison, and Strasser 2011, 161). While these deals vary from contract to contract, in their most complete form they cover records, touring, publishing, endorsements, merchandise, books, movies, and fan clubs—everything (Donnelly 2010). In effect, these contracts act as a form of horizontal integration as they aim to take over the roles of other parts of the music industry. Where musicians have depended on record labels for cash advances to record albums, everything beyond the sale of recordings and the associated copyrights of that music was left to the musicians. This means that whereas in the past touring, product licensing, and so on were part of the musicians' revenue, record labels are increasingly demanding a piece of that pie too.

I want to be very clear that the major record labels make a logical jump here. Their first position is that recording artists are being victimized by file sharers "stealing" music. From there the labels are jumping to the position that their revenues are down because of file sharing. Record labels then argue that one of the ways for the recording industry to remain viable is by pulling in revenue through their artists from other parts of the music industry. Ideology allows for the smooth functioning of this argument because the public has been led to believe that intellectual property is necessary and being violated by copyright pirates. If music fans want to keep listening to music, then the major record labels argue that there have to be fundamental changes in the legal apparatus and norms of the music industry. While directed to music consumers, this argument does its real work on the musicians; in other words, major record labels stipulate to new recording artists at the time of signing that business has changed.

Over the past decade, the recording industry's narrative of declining revenue from recorded music sales as a result of file sharing has been so pervasive that it is difficult to find a news article that does not position the 360 deal as a result of this decline. Whether it is industry representatives in *Billboard* claiming that "as a result of the significant decline in album sales, record labels are exploring new and drastically different business models" (Branch 2009), or rock critic Steve Knopper claiming that 360 deals make sense because "album sales [are] down twenty-five percent since 2000" (Knopper 2007). There is no reflection on the condition of the recording industry in the move to these deals.

In fact, following Branch's and Knopper's arguments, 360 deals seem to be the next logical step for a recording industry reeling from lost revenues because of declining CD sales. This position is erroneous for two reasons. First, it is problematic to contend that the recording industry suffered due to the digital transformation because, as I have been arguing throughout this book, the recording industry is in a stronger position now than it was in the 1990s. Second, the recording industry has removed the recording artist as victim from its rhetorical position and replaced it with record labels as the victim. There is no longer a need for the recording industry to appeal to its recording artists because the public already believes that the recording industry is in decline. The RIAA and the major record labels now contend that file sharing is harming the industry as a whole without mentioning the impact on their artists. As a result, record labels look to their own recording artists to extract more profit to supposedly keep record labels (and by extension music) solvent.

To an extent, 360 deals should not be the least bit surprising because capitalists are always trying to figure out ways to extract more surplus value from labor; however, the rise of these deals is a result of the recording industry's argument that recording artists are the victims of file sharers "stealing" their music. What I demonstrate above is that record labels are arguing that their recording artists are struggling to survive as a result of illegal file sharing; the argument goes that recording artists are losing revenue that they rightly deserve because all workers deserve the fruits of their labor. However, recording artists have always earned money from playing live, making appearances on TV or in movies, and selling merchandise at shows; record labels were not arguing that these sales were declining because of the availability of free music online. In fact, many recording artists were arguing that the opposite was the case: the availability of free music on P2P sites was drawing larger crowds to shows. In the Senate Judiciary Committee hearing, Roger McGuinn even argued that sales of music by his band, the Byrds, had increased through the band's website (Hatch 2000). The opportunity created by the Internet gives recording artists the ability to generate income through other streams. As music fans search for new ways to listen to music and musicians begin thinking about ways to profit from their fans' consumption, new media threaten to destabilize the position of the recorded commodity distributed by record labels.

The system under which recording artists are dependent on major record labels to produce, distribute, and promote their music has never been in the material interests of musicians, while live performance and merchandise sales have put food on these musicians' tables. For example, McGuinn hardly made a dime from fifteen albums that he recorded for Columbia Records. McGuinn explained,

> When you sign a contract that says you are going to get 15 percent, there are ways the record companies have of not paying these royalties. And in my experience, even though we have had number one hits with "Mr. Tambourine Man" and "Turn, Turn, Turn," I saw nothing but the advance, which is divided five ways. It was only a few thousand dollars apiece. (Hatch 2000)

As McGuinn explains here, recording artists do not see any revenue from their recordings unless they make back their initial advance out of their percentage of the royalties. Most recording artists pay themselves some money during the recording of their albums from the advance, but this rarely exceeds a few thousand dollars. Rather, it is likely that over the forty years that McGuinn spent in the recording industry, the bulk of his income came from performing live. The goal of 360 deals is to colonize the last vestige of autonomy that recording artists have with regard to revenue from their labor.

While recording artists have always been able to rely on their tour revenue to support them, 360 deals take away that revenue and attach it back to the initial advance. Under the terms of one of these deals, all artist royalties and income go back to recouping the initial advance. In a biting critique of these deals, "Buyer Beware" (2010), Bob Donnelly contends that the 360 deals are a result of record labels claiming not to earn enough money from recording revenues alone. He explains,

> What's more, record companies love to "cross-collateralize," a 31-point Scrabble word that refers to the practice of taking an artist's positive earnings from one category (e.g., publishing income) and applying it as a record company expense that affects the artist's unrecouped balance in another category (e.g., the record royalty account). In other words, the labels are postponing the day when the act actually receives a positive cash flow from its end of the pipeline. Yet when it comes to the income that they would like to receive from an artist's 360 income streams, the labels want to keep 100% of the money they are entitled to, without applying (i.e., cross-collateralizing) any of it to reduce the artist's debt to the record company. (Donnelly 2010)

By cross-collateralizing, record labels ensure that they recoup a greater percentage of recording artist advances at the expense of the musicians themselves. This may make sense for some artists who receive a larger advance to help with tour support and whose managers are generally incapable of putting together a tour, but for the vast majority of acts, cross-collateralization means the same amount of money for fewer alternative revenue sources. The "bad news for new artists is that these deals can take the form of 'passive participation.' The labels make no additional advance payments and do no additional work; they just take a piece" (Gordon 2008, 13). While McGuinn and the Byrds never fully recouped any of the album advances, on a 360 deal the Byrds would have been held accountable for recouping their albums through tour revenue. Since record labels are pushing new artists to sign 360 deals, this points to the fact that they are unconcerned with their recording artists' ability to meet their material needs.

It is only through the rhetoric of an ailing recording industry that labels have been successful at forcing artists to sign 360 deals because the piracy panic narrative constructs an ideological position that naturalizes the major record labels' position. While the Internet provided musicians with the tools to subvert record labels, the digital transformation delivered record labels the ideological position to extract more value from their recording artists. Without the empathy on the part of music fans for their favorite musicians, the major record labels could not have constructed a believable narrative about the recording industry crisis because few people empathize with large corporations. Since the Internet creates new revenues for musicians, record labels have developed new ways to extract profit from those new revenues.

CONCLUSION

While some critics initially thought that the digital transformation of the recording industry would lead to a growing independence of musicians from record labels, the opposite has been the case. Record labels have seized the opportunity created by a weak public understanding of the way that P2P file sharing operates and an equally weak understanding of copyright law to demonstrate that recording artists were the victims of fans stealing music. In turn, record labels used this murkiness to establish new ways of extracting profit from their recording artists because even people highly critical of the recording industry believe that (1) it is struggling and (2) music fans need the recording industry in order to consume music. Record labels have expressed the idea that recording artists are victims and have developed an ideology that obscures the real relations of production; this allows the major record labels to implement recording contracts that further exploit the labor of their recording artists. Through 360 deals, record labels have used the ideological position that piracy is hurting the recording industry to colonize money from their recording artists' other revenue streams (i.e., touring, appearances, and merchandising) that these labels never had access to before.

Part IV

Digital Distribution and Surveillance

Chapter Seven

Distribution Then and Now

Similar to large corporations in other cultural industries, major record labels have maintained their dominance in the music industry largely because of their control of distribution networks—in particular, the access of major record label artists to brick-and-mortar stores.[1] As Nicholas Garnham and Fred Inglis (1990) explain in relation to the movie industry, the availability of media content is severely limited by distribution networks. While the cost of reproduction of media content is generally low, access to extensive distribution networks is limited and can be quite expensive. The size and cost of these distribution networks, Garnham and Inglis argue, creates monopolies and oligopolies for media corporations and allows them to exert influence over the content distributed on these networks. In the recording industry, small record labels and independent artists have historically had a difficult time getting their albums to record stores and outlet centers across a wide geographic area, but major record labels have developed the networks, relationships, and infrastructure to distribute their artists in a near-universal manner.

However, as music consumers turned to online sources for their music in the 1990s, the major record labels lacked a way to profit from selling music to consumers, and consumers turned to legally questionable sources for digital music. For this reason, disintermediation was a threat to the physical distribution system that worked to the majors labels' advantage. How did the recording industry address the decreased significance of physical distribution networks after the expansion of digital recorded music? In an attempt to regain control in the market, I posit, the major record labels turned to an intensive strategy of online surveillance to adapt to the challenge of disintermediation. This surveillance of consumers took two main forms: first, the industry tracked unauthorized file sharing in order to discipline listeners to use online stores to avoid lawsuits, and second, the industry engaged in extensive commercial surveillance to track listeners' consumption habits. Surveillance of online music consumption encourages music fans to consume music from online retailers and re-creates the advantages of physical distribution for the major record labels.

I define surveillance as the monitoring of human behavior with the goal of altering it (Foucault 1977) and explore how dominant firms in the recording industry engaged in strategies of legal and commercial surveillance to discipline consumers in particular ways. To this end, chapters 7 and 8 show the material effects of the digital shift in the means of music distribution. These chapters discuss why the digital system that has emerged benefits the major record labels. Contrary to recording industry statements about the inevitability of digital music distribution, I argue that this shift was caused by a dual surveillance of digital networks by both the RIAA and the major record labels.

PHYSICAL DISTRIBUTION

In 1995, the primary way that music fans could purchase music was at brick-and-mortar retail stores. By this time, compact discs were already the main media format of recorded music, while tape cassettes still constituted about one-third of overall music sold in the United States (IFPI 1996). Delivering music in physical media formats at brick-and-mortar stores is a labor-intensive process, and the costs of the physical medium helped to inflate the price of music. Not only did physical distribution increase the price of music for consumers, but it also made it difficult for small labels and independent artists to get their music widely distributed because of the high costs associated with producing, marketing, warehousing, and selling physical CDs. Furthermore, the major record labels have long owned the largest music distributors in the United States.

When a CD sits on the store shelf, it has already incurred a number of costs that are incorporated into the cost of the CD; see figure 7.1 for a percentage breakdown of the costs that contribute to the price of a CD. The price to manufacture and package each CD ranges from one to two dollars, depending on the number of CDs being printed and the type of packaging (Hull, Hutchison, and Strasser 2011). On top of the cost to make the CD is the actual cost to get the CD from the manufacturer to the retailer. Most distributors charge about $1.70 for each CD shipped (Hull, Hutchison, and Strasser 2011, 255). Additionally, retail stores have their own costs and profits that they work into the cost of a CD. A standard CD is sold for twelve dollars wholesale, but the retail price is often around seventeen dollars; this means that approximately five dollars of the price of a CD is a surcharge for the ability to purchase a CD at a store (Avalon 1998). The largest part of the cost of a CD is in the gross margin of the record labels, which can be from $5.51 to $6.86 depending on the rate of artist royalties (Hull, Hutchison, and Strasser 2011). Understanding the costs that go into the price of a CD is important because it is necessary to reflect on what contributes to the value of the CD. This is not to say that some costs should not be higher. For instance, it is likely that the record store clerk makes slightly above minimum wage—raising the value of the retail price *may* contribute to higher wages for the clerk.

By looking at the costs of selling a CD, we can begin to see how surplus value is extracted from the CD in a physical distribution chain. Marx's labor theory of value contends that all value is derived from the labor time that is put into producing a commodity.[2] The longer it takes to produce a commodity, the higher we can expect the price of the commodity to be because "all labor is an expenditure of human labor-power, in the physiological sense, and it is in this quality of being equal, or abstract, human labor that it forms the value of commodities" (Marx 1992, 137). This is no different in the music industry than in any other field of production. At one end is the labor of the musicians in writing and performing the music, along with the producers and sound technicians helping to record the music; at the other end are store clerks selling CDs to consumers, and in the middle, all the people who do manufacturing and distribution. Each of these workers sells his or her

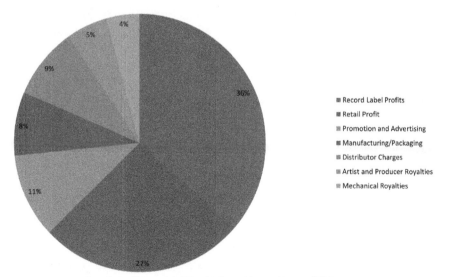

Data obtained from Park (2007) and Hull, Hutchison, and Strasser (2011)

Figure 7.1. Costs Contributing to the Price of a CD

labor power based on the socially necessary amount that reproduces the worker (Marx 1992, 274). While each worker may work the same amount of time, their labor does not necessarily cost the same amount because, for instance, it takes a lot of training and practice to replace a musician whereas the low-skill labor of the clerk is easily replaceable. Regardless of the value of each type of labor that goes into producing the commodity, under a physically mediated distribution system, every point in the production and distribution chain increases the end commodity exchange value as labor power is embodied in the commodity.

All of this is important because it increases the cost of a recorded commodity. Since the recording industry's commodity is recorded content and not the medium that recorded content is sold on, the production of an album or song only includes the costs that go into recording it. Manufacturing, distribution, and retail are costs that are beyond the cost of record production, and these costs are directly included in the price of a CD, tape, or LP. The actual cost of producing an album is contained in the artist's and producer's royalty portion of the price of a CD. Artists and their producers on a major record label generally net between one and three dollars per CD sold. The artist's portion of these royalties goes back to the record label to recoup the advance used to produce and market the album. Therefore, the actual cost of the music is contained within the artist's royalties. Every other cost that contributes to the price of the CD is nonessential for the production of recorded music. The easiest way for corporations to reduce the production costs of CDs is to reduce the amount that they pay for labor or eliminate the labor altogether. As I show later in this chapter, disintermediation, caused by the digital music commodity, has the effect of reducing the amount of labor that goes into producing and distributing recorded music commodities.

All these additional costs in physical reproduction and distribution raise the price of CDs and potentially give consumers less money to spend on more music, but this is a system that suppresses the ability of new bands and musicians to enter the music market. In the music industry, the major record labels control the means of distribution; by means of distribution I mean all the processes that pass music to consumers—David Hesmondhalgh (2007) calls these processes "circulation" (i.e., marketing, distribution, publicity, and transmission).

Most importantly, all the major record labels own distributors—for example, Sony Music Entertainment has a corresponding distributor called Sony Distribution (Hull 2004). These major record labels' distributors keep in constant contact with large individual stores and "one-stop" distributors (smaller distributors that mainly provide catalogs to mom-and-pop stores) to make sure that their artists are sold in retail stores nationwide (Hull 2004). Even if an independent artist contracts with an independent distributor or "one-stop," independent artists cannot guarantee that retailers will carry their music because there are larger profit margins on the major record labels' music; retailers want to stock shelves with music that will sell and will make large profits. The main problem in physical distribution lies in the number of retail outlets where music is available throughout the United States because "where geography demands of the music industry a highly developed and very mobile distribution infrastructure, control was with those who owned and operated the infrastructure of distribution" (Jones 2002, 217). Controlling the means of distribution has allowed the major record labels to have oligopolistic power over the consumption of music by restricting what music is distributed to retail outlets.

Similar to the changes that have occurred to the music commodity, the recording industry has shifted the primary retail stores over the past two decades. There have been three overlapping phases in relationships between major record distributors, musicians, and retail outlets in the physical distribution and sale of music over the fifteen years that this project explores (1995–2010). First, in 1995, mom-and-pop record stores were still very prevalent. Mom-and-pop record stores have always been known for allowing local and independent bands to place albums on their shelves, but it was difficult for a band to go to a large number of mom-and-pop stores to get their music in a substantial number of stores. For the most part, mom-and-pop record stores either were bought out by chain retailers or went out of business while trying to compete with large corporations; today, there are very few small independent record stores left. In the second phase, large record retailing chains, like Sam Goody's or Tower Records, began to put mom-and-pop stores out of business. These stores in rare instances allowed local musicians to put music on their shelves, but they were more selective. If a band could convince a chain to carry their album, the chain would sometimes carry it regionally or nationally, but usually the band was able to shelve the album only at the individual store that it convinced to carry the album. Some of these large retailers have gone out of business as a result of competition with big-box stores and digital distribution.

Third, the mid-1990s saw the rise of the big-box store (e.g., Walmart, Target, and Best Buy). Big-box stores changed the distribution and retail process for music sales by the year 2000 because their main business was not in music sales. Big-box stores had very limited space for music but sold what they had in very large quantities (Knopper 2012). One phenomenon that big-box stores developed is the concept of loss leaders. Loss leaders are sales that the big-box stores lose money on, selling below retail and sometimes even wholesale value in order to get customers in the store (Hull, Hutchison, and Strasser 2011). Big-box stores used music as a loss leader in the 2000s because it induced customers to come in the store to buy cheap CDs. Big-box retailers hoped that the customer would buy other, more profitable products while at the store. In this way, while record stores generated profit primarily on the sale of music and music related merchandise, music was only a small part of a big-box store's business plan.

Shelf space in the 1990s (as it remains today) was money for a retail store, and record store managers wanted to ensure that they had their shelves stocked with music that was selling (according to *Billboard* charts). For an independent band or musician trying to get their music in record stores in the 1990s, it was nearly impossible to get much distribution, but the major record labels' distributors could do this with ease. Distributors for the major record labels did business directly with large corporate retailers and could maintain contact with smaller stores through catalogs and regional salespersons (Hull, Hutchison, and Strasser 2011). At the other end of the spectrum, independent bands and artists had a difficult time contacting stores and convincing them to stock their music. In the 1990s, mom-and-pop record stores saw value in allowing local artists to put their music on the shelves—in fact, many mom-and-pop stores had a "local" section devoted to independent musicians from the area. Since mom-and-pop record stores shared structural similarities with local musicians (both usually shared a small-business mentality), they wanted to encourage space for these musicians in their stores. However, corporate chains pursued uniformity on their shelves, and with their larger scale, they had the data to support the type of music that would sell the most among a given demographic. To that end, in the 1990s record store chains were less likely to give local musicians the opportunity to sell their music in the store (Park 2007). Furthermore, corporate standards tended to regulate what music could be sold at record store chains, but sometimes a manager would stock music if they were interested in the musicians. Finally, big-box stores did not allow local musicians to put their music on the store shelves in the 1990s (Park 2007). Big-box stores shelved only the best-selling music at any given time (Christman 2002), and some stores (like Walmart) refused to stock music unacceptable to the store's brand image (Cloonan and Garofalo 2003).

Since the cost of producing an album has been relatively low for the past two decades due to increasingly cheap home-recording equipment, distribution has become the primary means of production that creates barriers for capital in the recording industry. The means of distribution incorporates not only the actual cost of shipping albums but also the cost of creating and running the system of relationships that connects artists to retailers. It therefore takes not only machinery and labor to ship albums from a CD pressing plant to retailers but also people who maintain contact (often through the availability of retail catalogs) with retailers to make sure that they have merchandise to sell.[3] In multiple ways, the recording industry "concentrated on developing the means of moving music from the point of manufacture (which in most cases it did control) to the point of sale (which in most cases it did not control)" (Jones 2002, 217). Through controlling these vast networks, record labels have had a distinct advantage over independent musicians and labels in providing music to consumers.

For example, even if a band were to bypass the major record labels (and their affiliate independent labels) by obtaining a small-business loan to cover the costs of recording, manufacturing, and marketing an album, without a distribution deal with one of the major record labels, it is nearly impossible for that band to get its music in stores. From the perspective of the retailer, it is better to work with established distributors (i.e., the major record labels) because they know that they will receive a steady flow of inventory, the major record labels will be marketing the music extensively, and the major record labels will buy back unsold inventory (Hull 2004). Maintaining this system was beneficial for the major record labels, even in spite of the five dollars that retailers were making on the sale of every CD, because it ensured that the recording industry remained structured the same way it had been since the 1950s. The momentum and stability of this system allowed capitalism to continue with this type of distribution despite the possibility of greater profits with a more efficient system of distribution.

MEDIA MEASUREMENT

When *Billboard* adopted Nielsen's SoundScan media measurement system in 1992, it was a groundbreaking change that linked *Billboard* charts with actual sales. This change was a product of digital media and fast communication networks. Prior to 1992, determining which albums were selling in record stores was nearly impossible because of the dispersed system of record store retailers, the vast majority of which were independent mom-and-pop stores. Record labels did not report the number of albums they sell, but rather reported their data by album shipments—the amount of albums shipped by a record label,[4] which has only a weak correspondence with actual album sales. Before SoundScan, *Billboard* would survey specific record stores in particular markets to try to obtain a broad range of data about what was selling; *Billboard* justified this method with statistical analysis, which demonstrated their limited survey data could apply nationally. This system of measurement benefited the major record labels because their artists were the only ones that could sell well nationally due to their well-developed systems of distribution and marketing. Furthermore, major record labels were aware of which stores were surveyed and would contact record store managers and clerks and give them promotional goods in exchange for high reports on specific albums (Watkins 2005). When *Billboard* conducted its surveys, they did not provide an assessment of what music was selling but rather demonstrated that certain albums were being promoted by major record labels. In the 1990s, sales data measured by *Billboard* through distribution networks helped to market recording artists on major record labels. However, information and communications technologies transformed the data collected in a way that destabilized previous notions about what music was most in demand.

When a musician had a top-selling album on the *Billboard* charts in the 1990s, the chart position increased radio airplay and helped the musician sell more albums. This created a cycle as more radio airplay helped sell albums, which raised this musician's position on the *Billboard* charts. This cycle helped major record label recording artists with nationwide distribution deals. "*Billboard*'s listings of the top-selling albums and singles determine who is recognized as the industry's leading performers, which in turn impacts radio and video airplay, sales, industry accolades, and finally, of all things, chart position" (Watkins 2005, 36)—or, as Marie Connolly and Alan Krueger contend, "being ranked high on the charts is important to artists because future sales and recording contracts are related to their placement on the charts" (2005, 30). When an album was perpetually at the top of the charts, radio stations and MTV were pressured by fans and the marketing departments of the major record labels to play songs from that album more often. In fact, the radio industry was so dependent on the charts that radio stations based their weekly meetings to design playlists around the arrival of music trade magazines (Rothenbuhler and McCourt 2006, 313). This reliance on the charts played to the advantage of major record labels with strong distribution and marketing because their artists were the ones always already in a position to perform well on the charts. Consumers were aware of and had access to music produced by the major record labels, which helped the position of those artists on the charts.

Since the major record labels were actively tampering with the *Billboard* chart system (by getting store clerks to forge surveys), many inside the music industry desired to create a more accurate system that relied on actual album and single sales. Despite opposition from the major record labels, SoundScan (owned by Nielsen) was adopted in 1992 as the album sales measurement on which the *Billboard* charts would be based. SoundScan uses a universal product code (UPC or barcode) to scan sales immediately into the system at the point of sale. With this system, computers (using information and communications technologies) report exactly what albums are selling at specific locations in a given week. "Whereas the previous system relied on the hunches of store personnel, the computerized methods used by SoundScan provided what the corporate world refers to as 'hard data'" (Watkins 2005, 37). When a record store clerk scans a barcode for a sale, if that store is participating in SoundScan, the information goes directly into the album rankings. Instead of relying on the perceptions of people working in record stores mixed with the knowledge of how many units are shipped, SoundScan enables a national picture of what music is actually being sold.

Point-of-sale data also enabled record labels to get information on their products in an instantaneous way. The implementation of point-of-sale systems allows record labels to get immediate feedback about what albums are selling and in what demographics. With more data on album sales, record labels were able to reduce the number of units that they shipped to record stores. Record labels could receive instantaneous information about what albums needed to be manufactured and shipped to retail stores instead of their general practice of shipping too many CDs and having to buy them back from stores. "Just in time" distribution is a production model that reduces excess inventory in the distribution process by better coordinating distribution and speeding up manufacture, so that only what can be sold is produced (Amin 1995). This "just in time" strategy reduces the need to spend money on warehouses and the overproduction of CDs that underperform. The SoundScan era of music retail created a more efficient system of delivering music to consumers by allowing record stores to respond to actual consumer demand instead of projections by record label staff.

The switch to SoundScan demonstrated two interesting phenomena in the recording industry: hip-hop and country music sold more, by far, than had previously been thought by the industry (Hull 2004; Miller 1999; Watkins 2005). This revelation quickly inspired major labels either to buy independent labels that specialized in country or hip-hop or, failing that, to work out distribution deals with these independents, thus increasing their profile within these lucrative markets. The industry also found out a lot about music consumers—for instance, that the largest buyers of hip-hop albums were white, suburban, middle-class teenagers (Watkins 2005). As a result, the content of both genres began to change as major record labels adapted the music to audience tastes (Watkins 2005). At the same time, although Sound-Scan had a profound impact on the genres of music that major record labels would produce over the next few years, it did little to affect the power and market share of the major record labels. If anything, as noted above, Sound-Scan led to *more* conglomeration as major record labels scrambled to incorporate independent labels that produced hip-hop or country.

Importantly, SoundScan data did more than just create accurate information about what music was selling; it also gave record labels information about who was buying music. Through surveillance, record labels became increasingly aware of their consumers. Record labels "can access national sales information by region, by specific Designated Market Area (DMA), by store, and more" (Hull 2004, 201). While demographic information is not available in these data, by knowing the location of the store and the time that the music was purchased, labels can reasonably assume certain information about their consumers (e.g., the individuals who bought 50 Cent's latest rap CD at Great Lakes Mall in suburban Cleveland were most likely middle-class white teenagers). In turn, this information could be used to help develop marketing strategies beyond stocking more music of a certain genre. It could mean that, given the data, labels encouraged their artists to tour where their music was selling the highest. Labels could even use these data to encourage marketing firms to use their artists for cross-promotional purposes. The fact is that UPCs and point-of-sale systems allowed more information to become visible about consumers than in a physically mediated means of distribution.

DISRUPTION IN DISTRIBUTION: THE RISE OF PEER-TO-PEER FILE SHARING

Following the development of the MP3 format (1995), the release of Windows 95, the widespread use of web browsers like Internet Explorer (1995) and Netscape (1994), and the increasing accessibility of high-speed broadband, it was only a matter of time before people would begin to exchange music over the Internet. When the first widely available peer-to-peer (P2P) program, Napster, became available in 1999, P2P users almost immediately began sharing music from their hard drives with other users. Even before Napster, there were sites that had made copyrighted music available online. In 1997 alone, the RIAA filed lawsuits against three sites that provided copyrighted music free online (Jeffrey 1997). One of these sites even encouraged visitors to the site to "share" music with others by uploading it onto the site, much to the dismay of the RIAA. File sharing has become ubiquitous with digital music, but the technology itself has no relationship to sharing copyrighted material.

Overall, P2P file-sharing programs and networks are tools that individuals use to exchange various types of files on their computers with one another. Users can directly view, access, and download shared files on other users' computers. While the use of P2P programs to share music was indeed a breakthrough, longstanding trends in technological development and in the consumption of music itself helped pave the way. First, the network technology developed by Shawn Fanning (Napster's founder) was not new in itself but rather a resourceful new way to aggregate data and give users access to a large number of files on other computers through search technology. Second, music fans have always shared music, from teaching friends how to play songs in folk cultures to the trading of mixed cassette tapes among hip-hop fans. Since P2P programs are part of a historical technological trajectory, they merely changed the speed and ease at which users could share media content.

Napster was part of a specific historic technological trajectory. It was a product of the Usenet system of distributed Internet discussion networks, which was established in 1980. Usenet used "newsgroups" that allowed users to access, read, and comment on articles from around the world. While the newsgroup has become obsolete, it did leave two distinct legacies on the Internet. First, newsgroups have become more generally known as blogging (derived from the term *web logging*), and today people subscribe to blogs with RSS feeds the same way that many people used to subscribe to newsgroups. Second, Usenet newsgroups created a way to share files that would soon become known as file sharing on P2P programs.

File-sharing networks like Napster work in a fairly straightforward way. First, ripping music from a CD onto a hard drive in MP3 format makes digital music easily transferable over the Internet (Burkart and McCourt 2006, 49). P2P programs allow users to find digital files (music being the most popular) on other users' computers. The architecture for each of these programs is different, but they all facilitate the trading of files between users. Napster, Kazaa, Morpheus, Gnutella, LimeWire, and other P2P programs give users the ability to access files on other computers instead of files on a central server.

P2P programs allow people to share files whether or not they have authorization to do so. The networks themselves do not discriminate, so some sharing may be straightforward piracy, while other forms of sharing are unquestionably legal (Zittrain 2008). For instance, I could choose to share everything that I have ever written on a P2P program, or an independent musician could share his or her music. To share files on a P2P program, users generally place the files that they want to allow other users to access in their "share" folders. Only the files placed in a share folder are accessible to other users—it is generally up to users whether they provide any files for others. In short, although some users share copious numbers of copyrighted files without permission, others share only files that they do have authorization to share. In effect, P2P programs allow a level of surveillance between P2P users by allowing them to view the share folders of other users.

Since the launch of Napster, the RIAA has claimed that P2P programs were intended for piracy only,[5] and they argued that P2P programs should be made illegal. As P2P technology use became widespread, the RIAA used the courts and provisions of the Digital Millennium Copyright Act (DMCA) to clamp down and restrict its use. Eventually, through these efforts, P2P file-sharing networks became associated, at least in the eyes of the courts and among Internet service providers (ISPs), with piracy, and most of these networks have either been closed down or restricted by ISPs fearful of RIAA lawsuits. The RIAA also claimed that this piracy was doing harm to its clients and that sharing files on P2P programs is equivalent to stealing from artists. By shutting down Napster, the Supreme Court sided with the RIAA's view that P2P programs existed only to violate copyright (i.e., "steal" music).

For their part, Napster's defenders argued, drawing on the Supreme Court's previous decision with regard to the VCR in the Betamax Case,[6] that just because the technology could be used to violate copyright, this was not enough to warrant restricting its use, given that the same technology had many lawful uses unrelated to piracy. Contrary to precedent, in *A&M Records, Inc vs. Napster*, the Ninth Circuit of Appeals ruled that Napster had to shut down because it was designed with the intent to circumvent copyright legislation (Langenderfer and Cook 2001); Kazaa received a court injunction for a similar reason. Neither ruling, however, precludes other P2P programs from being developed and operated. Napster was shut down because it used a central server that, the plaintiffs argued, could be used to stop the flow of copyrighted material. Kazaa, which did not use a central server, lost its case because developers had marketed it specifically as something that could circumvent copyrights. The discourse about P2P programs assumes illegality, but there has not been any ruling that all people who use file-sharing software are committing a crime.

By choosing to shut down P2P programs, the RIAA demonstrated that what was at stake was not the availability of music online but rather the structure of profit making in the recording industry. P2P file sharing opened up a system of distribution that no longer relied on intermediaries to get music to record stores across the United States. File sharing "also threatened to change the balance of power of who did the distributing" (Gillespie 2007, 43). Since P2P programs allowed anyone to use the Internet "as a public distribution platform, peer-to-peer systems challenged the recording industry's exclusive control over the distribution of sound recordings" (Burkart and McCourt 2006, 49). Through P2P programs, users became the distributors of music. In *Wired Shut*, Tarleton Gillespie argues that such user-to-user systems of distribution challenge

> the very assumptions the music industry relies on for profit and longevity: the privileged place of artists and their designated patrons; the assumption that cultural expression moves in one direction, from the few producers to the many consumers; the relentless commodification of that cultural expression; and the management of those commodities by established corporate interests through copyright. (2007, 43)

If the RIAA and the major record labels allowed P2P file sharing to continue, it would destabilize the very notions and ideologies that perpetuate consumer capitalism. Napster "was an ideological challenge to the very legal and economic principles on which [corporations] depend, and it did make plain some of the limitations of that system" (Gillespie 2007, 43). Allowing consumers to act as distributors would potentially develop new ideologies about the ownership and consumption of copyrighted materials—a development the RIAA was determined to stop. Ideologies about free media content available on the Internet ran counter to capitalism's process of commodification in the culture industry, a process that relies on the existence of copyrighted material.

At the same time, it is important to remember that the fast adoption of P2P programs by music fans was not merely the expression of a desire to steal music for free but in fact stemmed also from the recording industry's fear of, and inability to adapt to, new technology. In other words, P2P file sharing emerged partially as a result of the lack of availability of music online in the late 1990s. For years, commercial websites were not legally permitted to sell music because the major record labels would not grant the mechanical or performance licenses required to do so (Burkart and McCourt 2006; Park 2007), even as music fans were actively looking for ways to get digital music online.

In fact, the first website to make music available online for download was not a P2P program but rather an online retail store called MP3.com. MP3.com was a store that sold only the music of independent musicians and bands that authorized the distribution of their music, but it was ultimately shut down by the courts, after yet another RIAA lawsuit, because of its my. MP3.com cloud service.[7] My.MP3.com was an online digital locker system that allowed subscribers to access their music collection from any computer. To verify ownership of a song, subscribers had to place their CD in the CD-ROM drive of their computer, and the site would authorize access to that particular music on its server; users were not uploading the music from their CDs. In *UMG Recordings, Inc. v. MP3.com, Inc.*, Universal Music Group (UMG) argued that MP3.com was violating copyright law because it was making unauthorized commercial copies of copyrighted material, and the case was settled out of court. As a result of such aggressive tactics toward online distribution in the late 1990s, the RIAA pushed music fans directly toward Napster and other P2P file-sharing programs.

As new file-sharing programs have been created to get around the problems that previous services faced, the RIAA has found new arguments to shut these services down. There is a twofold problem with this approach. First, it pushes P2P programs further and further from the mainstream and endows these services with connotations of piracy and illegality. The appearance of illegality is one of the main ways that the RIAA shapes and restricts social behavior, akin to Michel Foucault's (1977) concept of governmentality.[8] Whether or not the state is involved, governmentality describes the ways people regulate each other's behavior.

In this way, P2P programs gain the connotation of being illegal, so people begin to avoid using them in order to avoid breaking the law; this is demonstrated by the precipitous decline of file sharing between June 2003 (when the RIAA announced its plans to file sharers) and November 2003, according to the Pew Internet and American Life Project (Madden and Lenhart 2004). The RIAA's success in shutting down P2P programs through the courts had the added benefit of associating the technology itself with the nefarious and the illegal. In a Pew survey from March 2004, 14 percent of adult Internet users admitted to file sharing but claimed to have stopped downloading music on P2P programs because of the RIAA lawsuits (Rainie et al. 2004). The Pew report shows that those who quit file sharing did so because of the lawsuits, and 60 percent of those who did not share files in 2004 would not begin downloading specifically because they thought it was illegal (Rainie et al. 2004). At the same time, not one file-sharing case had been tried before people quit using P2P programs; the appearance of the law superseded what was or was not determined to be legal.

Second, by constantly shutting down P2P programs, the RIAA has limited the ability of the networks to grow and evolve. Jonathan Zittrain (2008) provides an analysis of the ways that the Internet and PCs have moved from being open and generative to closed and proprietary; the "generative" nature of the Internet has always allowed users to tinker with what already exists on the Internet. The Internet and PCs have been generative because "they were designed to accept any contribution that followed a basic set of rules (either coded for a particular operating system, or respecting the protocols of the Internet)" (Zittrain 2008, 3). When people use programs, they begin to conceive of new ways to use them and better execute them. Part of what the generativity of the Internet creates is innovation. However, the RIAA's persecution of P2P programs has limited the number of people that use these tools and likely limited countless other uses that those people would have developed from tinkering with P2P programs.

P2P programs allow more than the sharing of files; they by default share personal information. When users log onto P2P programs, their IP addresses (or parts of their IP addresses) are available to anyone with moderate to advanced computer skills. A person's IP address can point to very specific personal information about that person. The RIAA and major record labels use this information to track down users, either to stop them from sharing music or to improve their own marketing and advertising strategies.

BRICK-AND-MORTAR RECORD STORES VS. ONLINE MUSIC DISTRIBUTION

When consumers walk into a brick-and-mortar record store, they face any number of options, from purchasing music to stealing music, from leaving empty handed to asking the clerk to make a special order. Any of these options is viable, and very little information remains about the person's visit beyond any transaction that may have been made while in the store. In the brick-and-mortar store, consumer surveillance, if present, is primarily geared toward preventing shoplifting. The clerk keeps a generally distracted eye on the store, convex mirrors may help the clerk see around blind spots, cameras may provide a permanent record that monitors shoplifting, and magnetic fields may lead a shoplifter to believe that an alarm will sound if she takes the merchandise through the doors.

All the same, these mechanisms have no way of monitoring what a person does after she leaves the store with her legally purchased album. She can go home, make a thousand copies (compact discs or cassettes) and give them away to all of her friends or sell them to people on the street. There is also no way to tell how much she listens to the music that she purchased (if at all). She may have purchased an album as a gift for someone else, rendering most demographic information about the purchaser useless, or she may have determined that she did not like the album and left it on her shelf to collect dust.

Consumer surveillance improved, as noted above, with the invention of Nielsen SoundScan in 1991, a system that allows the music industry to track the sale of specific albums using barcodes and point-of-sale technology. This tracking system allows record labels to get near-instantaneous feedback on the albums that are selling at any given record store. Still, although Sound-Scan gives an accurate picture of what is purchased, it fails to monitor how and by whom the music is consumed. There is little information available about the use of any music purchased by this customer. Furthermore, there is qualitative information missing altogether on this imaginary customer's trip to the record store. There are no data stored on what she specifically came to the store looking to purchase. If she is visiting the record store simply to browse, then she is limited by what is available at the store, and, more importantly, there is no way to track efficiently what she browsed compared to what she purchased. While she may ask a clerk for something specific that she has in mind or search a catalog, such fine-grained information on browsing habits is lost forever in the brick-and-mortar context.

Through the transformation of music from physical media to digital media, the surveillance of a person searching for music dramatically changes. With the digital availability of music, the actions of our imaginary record store customer can be observed in much more detail, as digital technologies radically increase the information collected about consumption habits. If she logs onto a P2P[9] program, her shared files are immediately available for viewing by other people on the network; this is even more visible with her playlists on Spotify, her purchases on iTunes, her "likes" on Facebook, her browsing on Amazon.com, her searches for bands on Google, and so on. First, the most visible bits of information about her are her genre preferences as articulated by the specific music that she possesses in her shared music folder. Second, there is a record of her search history. Third, there are data stored on the music she samples and the music she downloads. Finally, a computer-savvy person can identify personal characteristics about her from her zip code, her credit card purchases, and other websites she visits. The information obtained from her P2P program usage can then be paired with her preferences on other websites. It is possible to combine information from her P2P program with songs that she purchases from the iTunes Store or radio stations that she creates at Pandora.com. This information can be processed and used to produce and distribute music that is directly marketed toward her specific market segment, which can in turn be continually refocused toward her segment's changing "tastes." If it was very difficult to tell exactly which music interested our customer in the record store, it becomes very easy to monitor her music interests on the Internet.

MAINSTREAM DIGITAL DISTRIBUTION

Since online retail for music has become ubiquitous, it is important to look at how the online distribution of music has resulted in new avenues of revenue, profit, and control for the major record labels. One result of the proliferation of P2P programs was the jump-starting of the recording industry's search for a way to distribute music profitably online. The International Federation of Phonographic Industries (IFPI) consistently articulated the need for a "legitimate" way for consumers to consume music online throughout the early 2000s. While this concern for legitimate online music retailers always already delegitimizes the use of P2P programs, it also points to the lack of coordination between record labels to produce an alternative. Now there is a robust system of online music distribution that stretches beyond conventional means of consumption as the recording industry continually finds new ways to extract revenue and profit from recorded music. These emerging forms of distribution use different types of information and communication technologies, including the Internet, gaming systems, cable, mobile networks, and satellite radio.

While a number of online music retailers existed early on (such as MP3.com), they focused on independent music. It was not until the iTunes music store's launch in 2003 that major record labels fully supported the sale of digital music online. The iTunes Store allows users to buy AAC files online by song or by album. While other online retail services are available, iTunes remains by far the largest online music retailer and credits itself as the largest music store in the world—ever. In 2010, iTunes had the largest library of any of the online music stores, giving it a competitive edge over its competitors with over 13 million songs available for download ("Bits & Briefs" 2010). Other stores have niche markets,[10] but no other store can provide the same one-stop shopping for music as is available on iTunes. The size of its library makes iTunes the Walmart of online music stores (Baker 2007). If a person knows that iTunes has the largest number of songs available for download, he or she will go to iTunes first—it is not rational to begin searching for a song on a smaller site. Rather, users are likely to look at other sites only if they do not find what they are looking for on iTunes or if they are looking for specific niche genres. Users tend to download music from iTunes because it has the largest music library, and *iTunes* has become the colloquial or genericized term for digital music.

There are a variety of online music stores with different business models to choose from, but iTunes remains the largest and most popular source for online music. People frequently refer to new technologies by the best-known or first brand in a market. The recognition of a brand gives it cultural capital that can generally be transformed into economic capital. This can be seen with Xerox for copying machines or Napster for peer-to-peer file sharing, but Rosemary Coombe (1998) points to the complexity of the legal distinction between the brand's trademark and its popular usage. Legally "brand names and trademarks may become part of public discourse, but they do so in a narrow and positivist fashion that understands the public role of signifying forms to be purely referential" (Coombe 1998, 79). While companies work hard to maintain their trademark's distinction, they also benefit widely from their name's ubiquity in the popular lexicon. A combination of name recognition, ease of use, and library size has made iTunes synonymous with online music stores.

Aside from per song downloading, there are subscription services that allow users to download music for a monthly fee. After proclaiming that the recording industry is "dead," Jimmy Iovine, former CEO of Interscope Records, asserted that the only solution for the recording industry is digital "subscription. Without it, there is no business" (Fricke 2012, 61). While Part I demonstrates that this is clearly an overstatement, Iovine's claim points to the feeling within the industry that labels would prefer to profit from other forms of distribution than the iTunes model. And increasingly, subscription services, such as Iovine's Beats Music, which he created with Dr. Dre, are becoming the primary sites of music consumption. The fundamental idea behind subscription services is that music listeners will no longer own music; rather, they will rent it. This is another fundamental change in the nature of the recorded commodity. After the Ninth Circuit Court shut down Napster in July 2001, it reemerged in 2002 as a subscription service where users can download unlimited music per month as long as they pay their fee. As soon as a user fails to pay the fee (or cancels the subscription), and the computer or device that stores the music is connected to the Internet, the files cease to work because the use of subscription files is contingent on the payment of a fee. In 2003, Napster[11] arranged a deal with Penn State University through which students attending the university would get automatic subscriptions paid for by their technology fees ($160 at that time).[12] The idea behind this move on the university's part was to avoid policing their students for file sharing. The recording industry intended to expand these partnerships with universities as part of an antipiracy campaign (Madden 2009). In effect, fees at these schools go to paying for a music subscription, but upon graduation, students have nothing to show for this money paid into the system unless they continue to subscribe. If the schools were concerned about consumption habits of all students, they could just pool that money to make a number of free CDs available for each student to pick up every semester. There seems to be a very nefarious relationship between institutions of higher education and the recording industry. On a larger level, these subscriptions fully embody the idea that music consumers have gone from owners to users/renters, as they retain no material commodity as the result of payment already contributed after their music subscription ends.

In the past few years, more subscription services have surfaced, some of which are tied in with larger media conglomerates. For instance, in 2010, Sony launched a new subscription service called Music Unlimited that operates through its new computer software, Qriocity, which is built on the PlayStation Network digital store. Qriocity is an all-streaming service, whereas Napster allows users to download music and that music works until the user quits paying for the service. As of right now, Qriocity is only available at home and does not allow for mobile usage. The significance of Qriocity is that it allows people to stream music through the PlayStation 3, but users have to pay for other services to access the same music through other devices. By subscribing to music instead of buying music, music listeners change their relationship to capital because instead of having a durable good that they could resell (potentially for a higher price), they are paying for a service that they have no right to upon termination of the contract.

Where most music services that allow users to stream music are on a subscription basis or function like an online radio, Spotify allows users to stream whatever music they want to listen to for free for up to twenty hours per month. Spotify users can upgrade their account for $9.99 per month to stream unlimited amounts of music. To begin listening to music on Spotify, all that is required is a Facebook account. What is not so obvious is how musicians are paid for music consumed on this website. In addition to ad and subscription revenue, Spotify considers its users' data part of the payment for streamed songs. Spotify gives labels "a record, by location, age, and gender, of every single time a track is played" (Greeley 2011). The surveillance that Spotify presents is in many ways more specific and easier to access than the surveillance available through other big data monitoring sources. Instead of having to aggregate data through IP addresses, the information on Spotify is readily available because of its link to Facebook accounts. Patrick Burkart and Tom McCourt describe this type of surveillance as "customer-relationship management (CRM)" because it is "based on personalization systems, which seek to build brand loyalty by creating an online 'experience' tailored to customer preferences" (2006, 94). The fact that users have to log in to listen to music gives marketers direct access to information about users. Whereas other forms of online surveillance (discussed in the next chapter) do not know the precise person using a given computer, Spotify's login information directly specifies who is listening to what music. Spotify thus presents the major record labels with a seamless way to track consumers.

Major record labels want to expand the use of subscription services by linking music libraries to telecom service contracts. Currently, they are working on ways to get people to subscribe to music services through their broadband providers. The recording industry considers this type of service as "'bundling' music with other services or devices—be it an ISP subscription, a mobile phone or a portable player" (IFPI 2008, 15). Their assumption is that these subscriptions will bring nearly unlimited cash flows. In the United States, 82 million people have broadband service; if ISPs charged ten dollars per month for the service, record labels could net $820 million per month, or $9.84 billion per year. Billboard reports that if "30 million subscribers [pay] $5 per month, [then] the revenue to content owners would be $900 million" per year (Peoples 2010). The recording industry is "optimistic" about the idea that these subscription services could generate perpetual revenue at very little cost for the labels themselves (Peoples 2010), and these revenue streams are available through the digital media and networks that exist today.

As new technologies develop to keep people plugged into the Internet at all times, the major record labels have developed new ways to get music to consumers and have expanded consumption as a result—whether that involves working with consumer electronics companies to develop Internet radio that can stream into cars via wireless cell towers, or creating subscription music services for smartphones, or working with game companies to license music for new games, or placing music videos on YouTube, and so on. There has also been a move by the major record labels to develop apps that work with smartphones, tablet computers, and TV entertainment units (e.g., Xbox 360, PlayStation 3, or Roku) (Bruno 2010a). These apps create new ways for people to consume music: the labels are working on apps with subscription services that stream audio, webcasts, on-demand music, and more. Record labels have begun to release albums as or in apps with special content available to the consumer such as pictures, bios, and interviews similar to what would be available in CD jewel cases (Bruno 2010a). Many of the new apps available are supported by advertisements. Since apps are growing at such a fast rate in different formations, it is difficult to tell how much capital the major record labels are accumulating from them, but they are generating revenue in new ways that unit sales cannot account for. These apps are also highly capable of sharing information about their users to the app developers. This expands consumption partly through DRM that restricts the transferability of music from one device to another. In effect, consumers must pay a separate fee to have music as a ringtone, for example, even if they have that same music on their iPod.[13]

Through the above emerging forms of distribution that utilize new information and communications technologies, the recording industry has recreated the dominant position that it had as a result of complex physical distribution chains. While these new distribution methods are constantly changing, they all increase the ability of record labels to track demographic information about consumers. Surveillance is a critical component for major record labels to maintain the significance of copyrights and to collect comprehensive data about consumption.

Chapter Eight

Watching Music Consumption

The Internet dramatically changed the surveillance of music consumers.[1] The recording industry uses this new online surveillance to secure its dominant position within the music industry. There are two types of surveillance that the recording industry uses online: (1) the RIAA monitors P2P programs and files lawsuits or requests Internet service providers (ISPs) block users' Internet access; (2) major record labels monitor these same P2P programs, and online music consumption more broadly, to market their music more accurately to consumers. It is important to understand the dual characteristics of the process of online surveillance in order to analyze the way that large corporations (both within and outside of the music industry) reproduce the relations of production (i.e., the relationship between labor and capital) during the creation of new technologies. The two types of surveillance described here represent both a visible and a hidden surveillance based on the same technology. Without the visible surveillance of P2P programs, which deters users from using these networks, major record labels would not have had the hidden surveillance to monitor music consumption for market purposes. P2P programs became both the source of copyright violations and the source of a new type of market analysis. The RIAA and the major record labels use both forms of surveillance together to structure online consumption habits. Major record labels and the RIAA used surveillance to change the distribution system, rather than attempting to stop all types of online music distribution and maintaining the CD as the primary recorded music commodity.

REPRESSIVE SURVEILLANCE

First, the surveillance of P2P programs structures consumption habits by condemning piracy and encouraging users to self-regulate for fear of being monitored and punished by industry associations such as the RIAA. Downloading music using P2P file-sharing networks has become synonymous with music piracy. In the discourse about downloading music, very little room has been left to discuss the *legal* downloading of music. It is not my intention here to explore the nuances of copyright law, as I touched on this in Part II, but rather to call attention to the way people have been forced to change their habits because of the RIAA's intimidation through surveillance. The RIAA created spectacular events by suing thousands of people of all ages, from teenagers to grandparents, for millions of dollars in some cases, for downloading (or rather sharing) music online (Lessig 2004; McLeod 2005). In the process, the RIAA successfully threatened thousands of people with the loss of their life savings by presenting them with an unsavory option: either settle out of court (and stop sharing music) or risk paying up to $2,500 per shared song if one loses in court and thousands more in legal fees even if one wins. In this way, the RIAA intimidated defendants into early settlements by threatening to tie them up in court for more money than they can afford to pay (Knopper 2009; Park 2007). Even though the RIAA quit suing users in 2008, they still monitor P2P programs and report file sharing to ISPs; in turn, ISPs are required to warn file sharers to stop, and if the file sharers continue using P2P programs, they lose Internet access. Both RIAA lawsuits/settlements and ISP Internet access restrictions eliminate the due process of law in the interest of capitalism.

While the RIAA could not possibly sue every user on any given P2P program, the spectacle produced by the RIAA in the news media and courts acts as a deterrent for most people by convincing them that the cost of facing a lawsuit (however unlikely) exceeds the value of downloading free music. The panoptic effect of Internet surveillance, when combined with the publicity surrounding file-sharing lawsuits, has therefore created a situation where P2P file sharers are never quite certain if the RIAA is watching them, and this uncertainty and fear consequently structures their Internet behavior. To further produce a sense of omnipresence, the RIAA even messaged users on P2P programs warning them not to break the law because file sharing is equivalent to "stealing music" (Park 2007, 85); again, a perpetuation of the piracy panic narrative. In this way, P2P programs began to resemble, for many file sharers, an Orwellian telescreen reporting their habits back to Big Brother RIAA.

Not only did the monitoring of P2P programs create a hostile and menacing legal environment for those downloading music, but it also brought an increased consciousness to Internet users that they were, in fact, not anonymous.[2] They could be seen. And since potential file sharers never know if and when the RIAA is observing their P2P usage, the RIAA's surveillance acts as a panopticon. For Michel Foucault, panopticons function by a web of institutional surveillance within which the watched can never be sure when the watchers are looking. As a result, such mechanisms of surveillance structure the behavior of those being watched as if someone is always watching (Foucault 1977). One major flaw in applying Foucault's concept of the panopticon to Internet surveillance is that his hypothetical subjects in a prison have no way to remove themselves from the surveillance, while Internet users can leave their computers (Lyon 2007, 59). Mark Poster argues that a better term is the "superpanopticon, a system of surveillance without walls, windows, towers or guards" (1990, 93). While Internet users can walk away from their computers, unlike prisoners, the panopticon remains a useful theoretical tool to think about power asymmetries in the surveillance of the Internet.

There is ample evidence to demonstrate a link between RIAA lawsuits and a reduction in file sharing. One team of economists observed file sharers before, during, and after the start of RIAA lawsuits and came to the conclusion that "individuals have, to a very large extent, responded in the direction intended by RIAA" (Bhattacharjee et al. 2006, 111). In 2004, the year after the RIAA lawsuits began, "among Internet users who have never tried music downloading, 60% say the RIAA lawsuits would keep them from downloading music files in the future" (Rainie et al. 2004); this was in addition to the 14 percent of file sharers, discussed in chapter 7, who quit file sharing following the lawsuits. According to a 2005 Pew Internet and American Life survey, "Fear of legal recourse deterred most former downloaders" from using file-sharing programs (Madden and Rainie 2005). This survey found that the highest percentage of former file sharers, 27 percent, quit specifically because of the RIAA lawsuits (Madden and Rainie 2005). Whether or not the RIAA was in fact observing P2P users and whether or not their use was legal became secondary to many users' fear of being caught downloading music. The risk of being seen sharing files illegally, and subsequently being sued, drove a number of users away from P2P programs and onto online retail stores. According to a Pew Internet and American Life Project survey, "the percentage of Internet users who say they share files such as music, video, picture files or computer games with others online dropped from 28% in a June 2003 survey to 20% in the November-December survey" (Madden and Lenhart 2004). This survey included the time period when the RIAA began suing file sharers; additionally, the report shows that Kazaa usage dropped by 25 percent between November 2002 and November 2003 (Madden and Lenhart 2004). There is significant evidence to demonstrate a link between the dramatic decline of file sharing and the RIAA lawsuits against file sharers. The piracy panic narrative in the press led file sharers to believe that they were doing something illegal, which encouraged many people to quit file sharing—regardless of the actual law.

At the same time, simply deterring music consumers from sharing files was not enough to satisfy the recording industry. Consumers needed somewhere to go. In 2002, the International Federation of Phonographic Industry (IFPI) stated, "the industry's greatest challenge" was "developing a legitimate online music business, in the face of rampant internet piracy" (IFPI 2002, 2). And this is where online music stores such as iTunes come in. This IFPI report recognized that the recording industry had to develop a legitimate online business before they could convince music downloaders to quit sharing files.

> As long as internet pirate sites continue to thrive, the development of the legitimate business will be seriously hampered. In order to create space for a legitimate internet business to develop, IFPI and its affiliated national groups across the world will continue to aggressively fight Internet piracy. This is done through continuous monitoring of infringers, cooperation with ISPs and, where necessary, taking legal action. (IFPI 2002, 3)

The IFPI acknowledged a connection between stopping piracy and creating a "legitimate" alternative. In 2004, the *Recording Industry in Numbers* report acknowledged that fighting piracy and developing a legitimate online business model were two sides of the same coin (IFPI 2004); this is important because this is the report on 2003, the year that the iTunes Store was released. In fact, the timing of lawsuits and the launching of the iTunes Store[3] points to a desire by the RIAA and the major record labels to coerce people who had become accustomed to downloading music online into using an Internet store. Steve Knopper directly links the opening of the iTunes Store with the attempt, through lawsuits, by the RIAA "to obliterate the millions of online pirates who showed no signs of breaking file-sharing habits" (2009, 182). The release of the iTunes Store also helped gain support from Hilary Rosen, then chairperson of the RIAA, and Roger Ames, then CEO of Warner Music, to file lawsuits because they thought "it was totally unfair to sue customers for downloading free music when they had no legal way to pay for it online" (Knopper 2009, 185–86). This, in turn, enabled major record labels, whose interests the RIAA protects, to profit from the new technology. Facing the legal gray area of P2P programs, consumers now had, in iTunes, a reasonable alternative. At the same time, their acquiescence came at a price—the abandonment of the many *legal* uses of P2P programs.

Getting P2P users to think that the RIAA was always watching was therefore one important way that record labels used surveillance of the Internet to ensure their production and distribution advantages continued on the Internet. To review the argument thus far: The disintermediation of the Internet presents risks to the record labels because it destroys the advantage that the major record labels have had over independent labels and artists in the production, storage, and distribution of music. Without an online music retailer, these labels would be forced to compete with a greater number of independent artists, and competition, even if celebrated by the true believers of neoliberal economic philosophy, is anathema to actual existing corporations (Harvey 2005). This form of surveillance had the effect of herding unruly consumers into a more amenable cage—iTunes.

MARKET SURVEILLANCE

The second method of surveillance is a more devious technique used by individual record labels to measure where there is demand for their music. While it was difficult for record labels to determine the demographic information of consumers in a physically mediated retail environment, the Internet created an environment in which demographic information about consumers can be known by monitoring their Internet consumption patterns. SoundScan made it slightly easier for record labels to monitor the sales of albums at retail stores, but the data collection was limited to purchases. With the Internet, the recording industry can see not only what is being sold in online and offline stores but also what is being downloaded on P2P programs, what music is being streamed on Internet radio stations, what sites users are visiting, what people are searching for on music sites, how long they are viewing these sites, and where site visitors are coming from and going to before and after they visit a site. Mark Andrejevic describes this surveillance as a by-product of the *"digital enclosure*—the creation of an interactive realm wherein every action and transaction generates information about itself" (2007, 2). When we enter the digital enclosure, we provide corporations with massive amounts of information about who we are, what we buy, where we live, and more. Andrejevic also uses the term to draw a parallel to feudal land enclosures, which created classes around access to the means of production. "A similar division can be discerned in the emerging digital enclosure between those who control privatized interactive spaces (virtual or otherwise), and those who submit to particular forms of monitoring in order to gain access to goods, services, and conveniences" (Andrejevic 2007, 3). Internet users give up their data for access to websites, in this case music websites. The composite data collected on Internet music listeners can be used in a number of ways to generate profit for leading recording companies.

The market surveillance of music listening habits on the Internet allows for the music industry to shift its strategies for selling music. Over the past several years, major record labels have become interested in services such as those provided by BigChampagne, a company dedicated to tracking demographic information from P2P programs, online retailers, and Internet steaming and subscription services to provide data about who is downloading particular songs online. A 2003 *Wired* article explains that "by matching partial IP addresses to zip codes, the firm's software creates a real-time map of music downloading" (Howe 2003). Let me stress that while the RIAA is not directly suing users anymore (Reardon and Sandoval 2009), BigChampagne was providing this information to the major record labels during the zenith of music piracy lawsuits. In this way, at the same time that the major record labels "are doggedly fighting to prevent online piracy, these companies are monitoring file sharing and selling that information to the record companies for a hefty price. WebSpins and BigChampagne monitor what Internet users are sharing on peer-to-peer fileswapping services" (Lawrence 2004, 30). Fighting file sharing and the commercial measurement of P2P programs have existed concurrently since the first lawsuits were filed; they constitute a dialectic. There would not be lawsuits without the same capacity to analyze the music market.

Moreover, it is equally important to mention that BigChampagne is not intended for use by everyone. Its services are tailored to and priced for high-end commercial clients. The cost of using BigChampagne is anywhere from two thousand dollars per month to track one album to forty thousand dollars per month for a record label to monitor its entire market. BigChampagne's scope of services changes depending on the money customers have to spend. This is important because it means that only corporations heavily endowed with capital can access and utilize data from BigChampagne while independent artists and labels cannot afford this information. Since only large corporations with significant capital can afford BigChampagne's services, the cost to use BigChampagne creates a structural advantage for major record labels. Knowledge is power, and the more corporations know about their consumers, the more effectively they use that knowledge to generate new streams of revenue. This is a critical part of how media corporations more generally have maintained their dominant position in the culture industry despite widely available free content online.

Data collected by BigChampagne are important for the major record labels because they can give them insight into where exactly, and with whom, a certain artist is popular. Since BigChampagne collects data across platforms (online retail, subscription services, web searches, streaming sites, band websites, etc.), these data are fairly extensive in scope. While brick-and-mortar retail stores can provide data on the consumption of specific genres and artists, this information is quite limited by the commodities available to purchase and the assumptions that record labels and distributors have about specific geographic demographics. As an article in *Wired* puts it, "Big-Champagne's music panopticon does a fine job of summing [complete data] up in a neat package to give insiders an easy way to see what's going on in their world" (Van Buskirk 2009). Major record labels have a near-perfect picture of who is buying what music, and this creates information on what music to record and where to sell it.

Furthermore, with BigChampagne's data, record labels can pressure radio stations in a particular region to play an artist's hit song by arguing that the music has been downloaded *x* number of times during the past week in that region. The record label can also use the data to encourage radio stations in a specific region to conduct on-air interviews or promos because of an artist's local popularity. Then record labels can use this location-specific data to make sure that an artist's tour passes through the regions in which that particular artist is most popular to play a concert.

Moreover, these data can in turn have an impact on how *other products* are marketed. Manufacturers of other products will hire a musician as a sponsor because they know that a similar consumer demographic buys the manufacturers' products and listens to a certain type of music. The following is an example of how this process works:

> Say a marketer has discovered that fans of the band Kings of Leon have a particular affinity for the beer brand she represents. According to the [data from August 2009], she should consider seeking out users of Last.fm, Napster, iTunes and Rhapsody, but not file sharing networks (represented by the Top-Swaps category) or Yahoo Music, where the song didn't make the top ten. (Van Buskirk 2009)

This type of market monitoring allows advertising synergy to function in increasingly parasitic ways not only by creating advertisements within advertisements but also by directly targeting people based on their personal consumption tastes, as inferred through their online data profile. As these marketing tactics become further refined, and the data profile becomes more specific, the data profile begins to create the subject that they seek to measure. Andrejevic says "the alienated world envisioned by interactive markers is one in which all of our actions (and the ways in which they are aggregated and sorted) are systematically turned back upon us by those who capture the data" (Andrejevic 2011, 85). People begin to consume the goods that marketers think they would like to consume because data lock people into specific circuits of consumption as alternatives are eliminated. Pandora radio provides a good example of how these circuits develop because as users create a radio station (through a thumbs up or thumbs down for different songs), Pandora uses an algorithm to predict other music that this user may like. As users enter more information, Pandora becomes more refined. Eventually, Pandora will no longer play new music outside of a user's very specific tastes for that station. Data from BigChampagne can then be used to target advertisements within this user's radio station based on the demographics of users with similar stations.

This said, not all the consequences of BigChampagne's digital enclosure (a virtual space where every move leaves a record of itself) are likely to be negative for artists. For instance, if the musicians become increasingly concerned with getting fans to their shows (as opposed to moving CDs out of Tower Records), then the recorded music becomes more of a promotional tool for the live product. While the major record labels' dominance in the surveillance of market data on the Internet still favors the interests of those artists linked to major record labels, it could help change the way that people conceive of the commodification of music. It could, in fact, signal a move away from the industry fetishization of the recorded product and back to the importance of live performance. However, the continued disciplining of consumers to purchase music online contradicts any move toward a music industry that is unconcerned with the monetization of the recorded product.

Although BigChampagne has revealed that major record labels use its services, the labels have tried to keep their usage private because the RIAA's lawsuits assert that every use of P2P programs is an act of piracy. The IFPI's reports are very explicit that file-sharing services are "illegitimate" (IFPI 1996; IFPI 2002). In effect, the fact that the major record labels use Big-Champagne to monitor P2P programs demonstrates that there is a *legitimate* reason to use P2P programs: to collect information on their users to market music better. Their use of BigChampagne thus contradicts the blanket condemnation and confers legitimacy on P2P programs. There is a contradiction in that individual users' Internet habits are always known, but corporations hide in the shadows of the Internet. In fact, this is one of the greatest contradictions of the Internet because, as Andrejevic asserts, corporations on the Internet hide behind a wall of privacy, while they have attempted to eliminate all semblances of privacy for their customers (2007, 3). Record labels are in a position of contradiction when they use P2P programs to obtain information about their consumers' consumption habits by using surveillance because of the inconsistency between RIAA lawsuits and BigChampagne. "If the labels acknowledge a legitimate use for P2P programs, it would undercut their case as well as their zero-tolerance stance" (Howe 2003). The litigation spectacle that creates the RIAA's panopticon is a thin veil that has protected the absolute interests of the major record labels: that is, to profit from the sale of music. But consumers always already know that they are participating in the commodification of music because the "spectacle is the stage at which the commodity has succeeded in totally colonizing social life. Commodification is not only visible, we no longer see anything else" (Debord 1994, 21). It is not that consumers fail to see that the major record labels are more directly in touch with their desires, but rather that consumers have begun to see this process as ubiquitous.

Furthermore, downloaders labor for the recording industry so that record labels can better market their product, while some of the same users are being prosecuted for downloading that product without paying for it. Downloaders in effect become "indirect knowledge workers" because while they create surplus value for the major record labels, they are not directly producing knowledge but rather produce the data that are used as knowledge (Fuchs 2010, 186). The act of consuming digital music creates an audience commodity that can be measured and used to generate profit. For Dallas Smythe (1981), the "Consciousness Industry"[4] sells audiences to advertisers. Instead of viewing media consumption as leisure time based on the "free lunch" argument that media corporations perpetuate, Smythe explains that audiences actually *work* by consuming media because the audience's watching of advertisements is what pays for programming. Media producers in turn create content that appeals to certain audiences that can be easily sold, based on demographics, to advertisers. Creating a specific type of audience can help generate more revenue because advertisers can be assured that they are connecting with their consumers. These data allow record labels to efficiently market artists' music to fans, and the data can also be sold to other companies for cross-promotional/advertising purposes. Through P2P technology, the major record labels can use both illegal and legal downloads as raw data to obtain an anticipatory view of new markets and structure the norms and laws that govern consumption patterns of music downloaders.

Market surveillance by BigChampagne is possible only because of the RIAA's opposition to the illegal downloading of music. RIAA lawsuits create a layer of panoptic protection for the RIAA hackers who are maliciously accessing P2P programs to litigate against downloaders of corporate music. Legal respect for intellectual property rights secures legal protection for these RIAA hackers. The RIAA claims their intrusions into P2P programs are legitimate because they are "fighting piracy." This makes it difficult for P2P program advocates to defend themselves on privacy grounds because the RIAA can easily twist a demand for privacy into an admission that P2P program advocates have something to hide. Since there is a lack of privacy on the Internet, corporations can obtain information on Internet users, and as the RIAA's panoptic surveillance has increasingly scared music listeners away from P2P programs, the infrastructure for that surveillance has cleared the way for BigChampagne to monitor Internet music consumption.

On a certain level, the recording industry must ideologically sell itself as the only legitimate source of music in order to allow its rhetoric about piracy to function. Music listeners are sold the idea that downloading free music could destroy the music industry, but the gap between the music *industry* and music as *culture* is rarely identified (i.e., music will exist long after its deindustrialization). The RIAA has been arguing that what will end music is the "illegal" use of Internet technologies, but what actually concerns the RIAA is the end of the major record labels' oligopoly.

The recording industry has alienated consumers by using surveillance to discipline them and delegitimize P2P technology. Yet, if the surveillance created by BigChampagne were broadly publicized, the industry's shell game would be exposed by illuminating the actual uses of P2P technology that even the recording industry cannot resist. Because the major players of the recording industry act separately in their respective forms of surveillance of the Internet and because only the RIAA's panopticon garners attention in the media, Internet surveillance functions to reinforce the major record labels' power in digital capitalism. If consumers were aware of this contradiction, then they might not be as docile in accepting the recording industry's conceptions of ideal Internet usage (then again, they may remain equally docile for any number of reasons). However, the recording industry's surveillance strategies allow current Internet use models to be so ubiquitous that the downloading of music through online retailers is now the standard model for music consumption.

Surveillance technology has created a situation in which the RIAA targets file sharers for illegally downloading music on P2P networks, while the major record labels have used the same technology to better market to their consumers. The result of this surveillance is that the major record labels reestablish their dominance within the larger music industry because the recording industry has deterred users from using P2P programs and only major record labels can afford BigChampagne's services.

RIPPLES OF MUSIC SURVEILLANCE

The major record labels' use of both the RIAA to sue P2P users and BigChampagne to monitor the market directly links two forms of surveillance. However, the recording industry has not stopped at the monitoring of digital music downloads on the Internet because digital music can be surveilled long after its initial download through computer code called digital rights management (DRM). DRM acts as an additional level of surveillance for corporations and causes music listeners to conform further to a digitized, postindustrial process of production. Record labels and online retail stores code music with DRM to limit the number of devices/computers a song can be played on, and DRM can track the movement of a song when the computer or device is connected to the Internet. This not only allows record labels to limit the flow of "free" music but also forces music listeners to change their music listening habits in very specific ways. As discussed in Part II, this coding, which consumers purchase with their music, in turn has an effect on how people interact with music in everyday life by precluding people from listening to their music in particular ways. DRM in music files is one instance of the way that capitalism is adapting to information and communication technologies to perpetuate the commodification of cultural media.

Digital downloads from online retailers complicate the social practices that have developed around the exchanging of music and in many ways end the ability of people to actually share (in the non-P2P sense) music with friends. Music fans create mix tapes of their favorite songs to demonstrate feelings and alert their friends to new music. The CD greatly increased the ease through which people could share music by allowing traders to burn multiple copies of the same mix without losing quality, an attribute that tape cassettes infamously lacked. Within some music subcultures, mixes act as a way for people to share new music with other members of their circle; this is especially true with hip-hop (Harrison 2006). In the case of hip-hop, most of the recordings shared in this way are underground (independent of label influence). At the same time, there is a risk that as the number of artists that rely on distributing their music using iTunes (and other DRM-encoded services) grows, the more difficult it will be for them to share music. While there are ways to get around DRM,[5] encoding music shifts the way people listen to new music as they can only make a finite number of copies from their original copy. Since DRM allows copyright owners to track how consumers listen to a song, music listeners have lost the ability to listen to music in privacy.

Until 2009, Apple's iTunes had one of the most restrictive forms of DRM embedded in the company's AAC files. The DRM in AAC files at different times restricted what type of devices a song could be played on or the number of devices on which the song could be stored. Initially, including DRM on AAC files was not Apple's idea but rather part of a deal that Steve Jobs negotiated with CEOs of the major record labels in exchange for licenses to sell music on the iTunes Store (Knopper 2009, 157–82). However, in addition to including DRM that restricted the number of machines on which an AAC file could be operated, Apple included DRM that restricted AAC files to iPods. While restricting iTunes to Apple products angered record executives, they accepted the restriction because AAC files included the industry's DRM (Knopper 2009, 179–80). At this stage, AAC files were aimed at reducing piracy by restricting the ability of users to share their files.

In 2007, iTunes began selling DRM-free music, iTunes Plus, for an added price, but only EMI's music was available in this format (Garrity 2007). After several years of negotiations with major record labels, in 2009 the iTunes Store began selling all music DRM-free (Lettice 2009). However, it is important to note that DRM-free music is not completely without DRM. Rather, DRM-free music no longer has the restrictions that used to be associated with iTunes, such as limitations on the number of machines that a song could be played on. DRM-free AAC files continue to contain information about the purchaser through metadata[6] and the basic code that restricts the appliances that can play AAC files. Since the RIAA and major record labels continue to assert that file sharing is illegal, they can use metadata located on songs downloaded from P2P programs as evidence that someone has been sharing files. The Pew Internet and American Life Project suggests, "Yet, for all the major changes in the industry's tactics, the relaxed attitude only goes so far. Through digital fingerprinting and other tracking technologies, the record labels are monitoring copyrighted content as closely as ever" (Madden 2009). By giving the iTunes Store's consumers greater autonomy in their use of AAC files, the recording industry is using other forms of DRM to monitor the depth and breadth of file sharing because now the RIAA can track the movement of files through P2P programs.

However, iTunes is not the only digital music service that is using DRM to restrict the use of digital media files. Many other platforms (e.g., subscription services) are developing what Tarleton Gillespie calls a "trusted system." In addition to encoding and decoding an encrypted file, a trusted system "also obeys a series of rules about the content's subsequent reproduction and redistribution of it, ensuring that such functions are made impossible for the average user" (Gillespie 2007, 52). While iTunes' AAC files before 2009 created the prototypical "trusted system," this system has become the standard for the recording industry. Ideally, a trusted system can force users to "pay-per-view" or restrict access to current customers (Gillespie 2007). Major record labels hope that by forcing consumers to pay for permission to listen to music, they can generate more revenue than in a physical media distribution system. This is only possible in a system where constant surveillance of consumption is in place. A trusted system is a surveillance apparatus that enforces parameters created by producers.

Surveillance has changed music listeners' listening habits because DRM has eliminated the ability of music listeners to share their music with other people. This not only changes habits; surveillance creates a type of discipline that structures the way that music listeners consume music. Whereas music consumers have been music hoarders since the invention of the gramophone, trusted systems are turning them strictly into music users or renters.

CONCLUSION

Since the Internet was created by the Department of Defense and developed further by academics, some theorists argue that it was devoid of capitalist intention (McChesney and Foster 2011; Rheingold 2000). As Internet users began seeking ways to distribute music files to each other, the concern was not for developing new ways to sell music but rather for finding new ways to share music with one another. This is not to say that capitalism was entirely divorced from the development of the Internet; certainly, there have always been people who intended to acquire profit from the Internet. However, as major record labels became aware of the potential of the Internet as a means of music distribution, there was a necessary move to develop online stores that could distribute their music online (keep in mind that record labels were also trying to develop a system before Napster even threatened their position). At the moment when alternatives to the recording industry's distribution arose before the industry had a chance to develop its own system, major record labels needed a way to stymie the proliferation of P2P file-sharing networks. By using the surveillance of the RIAA and BigChampagne, the major record labels were able to reduce the use of P2P programs, while using data from these networks to construct stronger marketing information.

While the major record labels were slow to acknowledge the effects digital music would have on the music industry (Knopper 2009), they more than made up for lost time with the aggressiveness with which they cornered the digital music market. Major record labels used lawsuits for intellectual property violations and employed media measurement companies to create a media environment that benefits the major record labels over independent record labels and independent artists. The tactic was to scare people away from using P2P programs while at the same time monitoring music downloaders' actions in order to know where to market music. After the realization of competition from independent artists on Napster (Arditi 2007), the major record labels responded with a strategy to reestablish their dominance in a new sales medium. David Park explains that the major record labels recreated their dominant power on the Internet with a four-prong approach:

> First, it wanted to give the public the option to purchase individual tracks and albums online, but at higher profit margins. Second, it wanted to offer some digital tracks for sale online before they were available in other formats. . . . Third, the industry wanted to create subscription services where consumers have to pay to access and purchase music and last, it wanted to divide music through various labels and exclusive content throughout a number of online distribution services. (Park 2007, 94)

These strategies have been the cornerstone for enticing consumers to buy more music online. Indeed, fans of particular artists will inevitably seek out bonus tracks by subscribing to any service if it means being able to hear more music by a particular artist. Again, these strategies lead to the expansion of music consumption. However, if the music is available somewhere online for free, that fan is likely to seek it there first.

Forcing customers to buy music from multiple online outlets seemed like a great idea in theory, but the major record labels' plan that Park outlines failed because too many online services developed between 1998 and 2003. Often these sites negotiated licensing agreements with one or two of the majors, which limited the music available on any given site; this is part of the reason why Steve Knopper claims that major record labels were willing to license their music to iTunes (Knopper 2009). Furthermore, without an intermediary to restrict access to the music market, consumers tend to get lost in different sites of consumption on the Internet; there is no substitute for the marketing departments of the major record labels, and as a result, marketing has become the intermediary for digital music that allows consumers/fans to find music. The accessibility of a wide range of online sites of consumption could have created more niche outlets for genres and multiplied those genres because there would have been less centralized control of distribution. This could have in turn created more room for competition from independent artists. However, it is easy for music consumers to get lost with so much music available online, and the marketing departments of the majors intervene to guide consumers to music.

Surveillance is a tool that the recording industry uses to maintain its dominance within the broader music industry. Together, BigChampagne and the RIAA lawsuits give major record labels a structural advantage over independent labels and artists. The exorbitant cost to use BigChampagne makes it only available to firms with significant capital; this means major record labels can use data to get a precise picture of the music market. Yet these data are not enough, on their own, for major record labels to maintain the distribution advantages created by brick-and-mortar stores; there is no substitute for the disciplinary apparatus that the RIAA has for monitoring and prosecuting P2P file sharers for piracy. It is the contradiction between the RIAA's antidownloading surveillance and BigChampagne's market surveillance that has re-created the dominance of the major record labels on the Internet—along with large marketing budgets. This strategy of surveillance is a clear example of how the recording industry has been an active agent of change during the digital transformation.

As the dominance of the major record labels continues, their position becomes less reliant on any one method of discipline because they have fundamentally changed the way music listeners consume music. DRM has allowed the major record labels to constantly track and control the use of their copyrighted material.[7] Along with the perpetual surveillance of music through DRM, the major record labels have effectively expanded the means of consumption by getting people to consume the same music in different paid forms. None of this would be possible without the disciplinary nature of the recording industry's surveillance methods on the Internet.

Most importantly, the recording industry played an active role in changing the distribution of recorded music. While major record labels did not anticipate P2P programs specifically, they did recognize the potential for digital distribution early and began to create a secure system of distribution. After P2P programs received the attention of the recording industry, the RIAA and major record labels employed surveillance to discipline music consumers to utilize the industry's preferred distribution network. However, the recording industry no longer preferred selling physical media (i.e., CDs) through brick-and-mortar retail stores; rather, the major record labels preferred to have music fans consume music online.

Conclusion

Compact disc sales have been declining for over a decade.[1] According to the recording industry, the decline in CD sales represents a larger crisis within the music industry. However, the strength of CD sales is not directly related to the state of the music industry, and there is little evidence to substantiate the recording industry's claims. This book has countered the piracy panic narrative constructed by the RIAA and reproduced by news media outlets, which leads many people to accept that the music industry is in danger of becoming unprofitable and thereby ending music production and consumption as we know it. Nevertheless, music production and consumption have profoundly changed because of alterations put forth by the major record labels themselves. These alterations in turn have strengthened the major record labels' power within the larger music and cultural industries.

Two comprehensive conclusions can be drawn from this book. First, contrary to the recording industry's claims, the major record labels appear to be in a stronger position financially and politically today than they were before the digital transformation of the recorded music commodity. For instance, as Part I demonstrates, the record labels have expanded the means of consumption through the utilization of publishing rights, performance rights, and synchronization licenses and created new digital commodities for consumers to buy. This destabilizes the convenient story created by the industry represented by the RIAA. Second, rather than being a passive victim of the digital transformation, the recording industry actively contributed to that transformation. As an example, the recording industry and other copyright industries were fast at work on the bill that would become the Digital Millennium Copyright Act (DMCA) long before digital music was a technologically practical reality.

The first conclusion scrutinizes the industry's commonly accepted piracy panic narrative by demonstrating that (1) the major record labels have adapted the relations of production to digital media and (2) the recording industry has expanded the means of consumption. As a result, the recording industry increased its ability to profit from technological change. Undoubtedly, digital music distributed via the Internet and other communication technologies is now the dominant medium for the recorded music commodity. Digital music has increased consumer access to the music commodity, and the changing nature of the music commodity has allowed the recording industry to profit multiple times from the sale of the same song to the same consumers. The major record labels expanded the means of consumption partly because digital rights management (DRM) does not permit music fans to consume digital music through multiple media formats (e.g., an MP3 cannot be used as a ringtone—consumers must purchase each file separately). Furthermore, record labels have begun to exploit publishing rights more than they did before the digital transformation. Finally, surveillance of consumption on the Internet has increased this revenue by alerting record labels when a song is streamed on a website. These changes have positioned major record labels to increase profits by increasing the quantity of unit sales and reducing the cost of each unit.

The second conclusion challenges the recording industry's argument that Napster surprised the industry and that everything that the industry has done has been in reaction to peer-to-peer (P2P) file sharing. If the major record labels were surprised by Napster, then we would expect to see that the industry was not developing strategies to deal with Internet distribution before the release of this pioneer file-sharing network in 1999. However, the recording industry began to change copyright law to adapt to the digital transformation, beginning with the Audio Home Recording Act (AHRA) of 1992, long before file-sharing programs and MP3s were developed. This book has examined four areas in which the recording industry actively altered the relations of production. First, the major record labels changed the fundamental nature of the recorded music commodity. Second, the RIAA was able to persuade the U.S. government to alter copyright law in order to reinforce the recording industry's business models in a digital media system. Third, the major record labels used the piracy panic narrative constructed by the RIAA to alter the structure of record contracts with their artists. Finally, both the RIAA and the major record labels have used surveillance to induce consumers to consume music through their digital distribution networks instead of through alternatives such as P2P networks.

The digital transformation of the recorded music commodity has not created new spaces for independent musicians to compete with major record labels. Rather, this transformation has reinforced the dominant position of the major record labels within the larger music industry. However, this should not come as a revelation because capitalism, as a mode of production, relentlessly pressures firms and industries to develop and use technologies to create a more efficient means of production. While there is no necessary connection between technology and this process, more often than not capital is in a position to take advantage of the state apparatus to reconfigure the relations of production. Internet distribution of the digital recorded music commodity allowed the major record labels to become more efficient by reducing the costs of both variable capital (workers along the production and distribution chains) and constant capital (the physical process of production and distribution, such as CD pressing plants, delivery trucks, storage facilities, CDs, etc.). Major record labels have used their dominant position in the music industry and their access to capital to re-create a distribution system and to develop a business model that maintains their dominant position.

IMPACT ON THE PUBLIC CULTURE

Unfortunately, the changes discussed throughout this book are not limited to the digital transformation of the music commodity. What is at stake is the protection of the public sphere and cultural production. Newspapers, television, film, libraries, book publishing, software, and more are affected by the changes to copyright law and the implementation of digital rights management. Ultimately, these policy and technological changes affect all of American culture and public life and occur in an undemocratic manner.

Culture is not something that any individual or group owns, but rather it is the process through which people make meaning of the world around them. There was a time when all culture was held in common. Following the development of copyright laws, ownership of particular cultural artifacts became assigned to individuals and groups. Over time, the boundaries between one cultural artifact and another has been strengthened by each change to copyright law. As the borders between cultural artifacts become stronger, it is increasingly difficult for people to play with culture. Play, experimentation, and improvisation are all important aspects of the cultural and artistic creative processes. However, the "wired shut" (Gillespie 2007) attributes of today's technological and legal apparatus limit the ability of cultural producers to play with previous works.

Hidden underneath the recording industry's piracy panic narrative is a strategy to control every use of a cultural commodity. This restructuring of cultural commodities shifts cultural ownership from the collective to the individual (by extension in the current system, the individual is the corporation). This leads to two consequences. First, it leads to the further commodification of culture. Since digital files with DRM can be tightly controlled, consumers are forced to purchase media multiple times. For example, if a person is subscribed to Netflix, they have unlimited access to live streaming of particular TV shows and movies until they unsubscribe. After this consumer ceases to subscribe to Netflix, they no longer have access to those TV shows and movies no matter how many times they watched a movie. In order to watch those same TV shows and movies, the consumer would have to buy the DVDs for which they have already paid. Sometimes this is a product of using multiple media forms, but technology is changing what constitutes a different media form. If someone saw a band at a concert, she was never entitled to a CD of the music she heard. However, if that same person bought a CD, she expected to be able to listen to it on her home stereo, her car stereo, her Discman, or her computer. On the other hand, if she purchased a digital file from iTunes, she would need to purchase a separate file to listen to it as a ringtone, for instance. People have to purchase media content on multiple formats to consume that content on multiple devices.

Second, the tighter control of copyrighted material in turn limits the production of new cultural artifacts. Borrowing has always been a part of cultural production, but in the digital era, borrowing becomes hyper-watched. For example, Prince requested that YouTube remove a video of a child dancing to a song of his because he claimed that it was an unauthorized use (Sandoval 2007). In the video, a baby is walking around playing with a toy, but Prince's "Let's Go Crazy" is playing in the background (barely audible). Upon hearing the music, the baby starts dancing. The parents wanted to show friends, and they were not trying to profit from Prince's copyrighted material. "Let's Go Crazy" is at most a cultural reference in the video that aids in the production of other culture. This is how culture is created; it is not piracy or theft. Unfortunately, the law and technology are being applied in ways that limit this type of fair use.

Not only are the current technological and legal changes to copyrighted material affecting fair use (a necessity of a democratic public sphere), they are being advanced through an undemocratic process. Both the multiparty copyright stakeholder negotiations (discussed in Part II) and the Napster hearing (Part III) are illustrative of the copyright policy-making process. In both cases, the public's interest was either excluded or supposedly represented through the public's position as consumers. By treating citizens as consumers, this process substitutes capitalism for democracy. Citizens must be actively involved in the policy-making process beyond their consumptive decisions. Whether or not a person can play with culture goes beyond their decision to purchase a commodity that allows for that play. The public needs to be an integral part in democratic discussions about changes to law, especially when those laws affect the public to the degree of copyright law.

While the recording industry set out to transform the music commodity by claiming that the economic interests of major record label recording artists were at stake, the RIAA fundamentally changed the way that culture is produced. Over a decade after the recording industry claimed to be dying, it is still going strong—possibly stronger than ever. Music sales are at an all-time high (exponentially higher than CDs at their peak), copyrighted music is ubiquitous (played in TV, commercials, sporting events, video games, etc.), and the industry is profiting from performance rights, publishing rights, and synchronization licenses. The piracy panic narrative allowed the recording industry to leverage Congress and the courts to create laws that would stop the flood of file sharing, and this has had the effect of locking down culture.

Notes

PREFACE

1. "Catalog sales are defined as sales of records that have been in the marketplace for over 18 months" (Hull, Hutchison, and Strasser 2011).

2. The author of the *Rolling Stone* article references such high-water marks for the recording industry as Miley Cyrus's 2013 Video Music Awards performance and Jay Z's free album, among other "strategic hot messes" (Hiatt 2013).

3. For example, as a way to save $300 million during the Polygram–Universal merger in 1998, executives proposed layoffs ("Top Managers Mull the State of the 'UniGram' Union" 1998); in the distribution division alone, Polygram proposed to lay off 130 workers and lower employee pay (Christman 1998). Ultimately, five hundred record label employees were laid off in the immediate aftermath of the Polygram–Universal merger ("500 Are Out in First Wave of US Layoffs" 1999). At the same time, according to Nielsen SoundScan, the merged conglomerate gained in overall market share, increasing from a combined 24.48 percent of the market in 1998 to 26.39 percent for the new Universal in 1999 (Nielsen 1999; Nielsen 2001). Following mergers, sustained revenue (by way of unit sales) mixed with layoffs and other reductions in overhead costs increases profit. The year in question is the same year that Sony and BMG music merged. That merger laid off more than two thousand workers in the United States (M. Newman 2005). The overall layoffs numbers included only those at the major labels; counting subsidiaries pushes that number higher as, for instance, Sony BMG closed Epic Records Nashville for an additional 20 jobs lost (these add up with more subsidiaries). The merger between Atlantic and Elektra (both subsidiaries of Warner Music Group) in 2004 resulted in the firing of 184 workers (Christman 2004); 170 workers were laid off at Arista in 2004 and an additional 150 workers in Sony BMG's distribution division in 2005 (M. Newman 2005).

4. In fact, the Recording Industry Association of America (RIAA), which certifies albums, changed its rules to certify *Magna Carta* platinum because previous rules stipulated an album must be on the market for a thirty-day period before it could be certified (Knopper 2013; Pham, Hampp, and Christman 2013).

INTRODUCTION

1. The RIAA is the trade association that represents major music companies' financial and business interests to the government and the American public. While it publicly claims to represent the entire music business, the RIAA focuses on the interests of the recording industry. Despite a broad membership base of a number of large and small record labels, the RIAA specifically supports the interests of the major record labels (currently, Warner Music Group, Sony Music, Universal Music, and EMI)—these labels and their subsidiaries dominate the RIAA's executive board.

2. *A&M Records v. Napster, Inc.* was decided by Judge Marilyn Patel of the United States District Court for the Northern District of California in May of 2000 (Patel 2000). The court decided in favor of the plaintiffs, which included most members of the RIAA. After appealing the ruling, Napster settled the suit in return for becoming a licensed subscription service for downloading music. Ultimately, Napster was purchased by Bertelsmann—a large media corporation and parent company of Bertelsmann Music Group (BMG), one of the major record labels.

3. Since the late 1990s, the predominant recorded music commodity has changed from the CD to digital files. Part I examines the way that the commodity has changed, how that change has affected music business models, and the way that this affects revenues and profits. The decline of CDs as the main music commodity has been paralleled by the meteoric rise of digital music. Unit sales are more than double what they were at the peak of CD sales (IFPI 2001; IFPI 2011). Furthermore, as I show, this change is the latest mediation in an industry where change has been constant since the advent of the gramophone/phonograph.

4. Over the past decade, the recording industry has shifted considerable effort toward utilizing performance rights to generate revenue. Global performance rights data include "licensing income from both radio and TV broadcasting and public performance—such as the use of music or videos in nightclubs, bars, restaurants and hotels" calculated by "local collecting societies, from both the use of sound recordings and music videos" (IFPI 2005, 20). In 2000, the RIAA did not track revenues from performance rights because the source was negligible (IFPI 2001). In 2004, the RIAA began tracking performance rights revenues and reported $9.19 million in revenue (IFPI 2005). By 2011, the RIAA was reporting over $130 million in performance rights revenue (IFPI 2012). There is a parallel rise on the global scale where the International Federation of Phonographic Industries (IFPI) reported close to one billion dollars in performance rights revenue in 2011 (IFPI 2012). This rise is the result of the expansion of the means of consumption as the major record labels have found more ways to generate revenue from their catalogs: from video games to television shows. At the same time, there are few upfront costs that go into producing this revenue because the music is already recorded and it does not require any printing or distribution. Additionally, the IFPI's reported data exclude "the performing artists' share" (IFPI 2005, 20); this is significantly different from the revenue totals reported from album and single sales, which include total revenue. As a result, most of performance rights revenue reported by IFPI is profit for the major record labels.

5. *Billboard* magazine is a recording industry trade magazine. It is known for its charts, but this magazine is what industry executives and "insiders" read to get an idea about new developments in the industry. On issues such as file sharing, it acts as a soundboard for the industry. However, news reporters for other outlets often read *Billboard* in researching their own articles. This creates a surprising willingness to retransmit major record labels' positions on file-sharing issues.

6. Noam Chomsky and Edward Herman highlight the way that information released by businesses and government appears as journalism in their "propaganda model" (Herman and Chomsky 2002).

7. Evidence of this decline is visible in both Nielsen SoundScan Reports and the IFPI's annual *Recording Industry in Numbers.*

8. While this book often presents details within a historical genealogy, it primarily covers the years from 1995 to 2010. The year 1995 is significant for several reasons, but most importantly, it is when the MP3 file was first released to the public. Before the development of

MP3s, digital music was only available in WAV files—a computer file similar in size to CD audio files. The large size of WAV files, combined with the small size of hard drives and slow Internet connections, made it difficult to use digital music on computers. Additionally, Windows 95 made PCs easier to operate, and Internet Explorer simplified Web browsing. The technological and legal advances made during 1995 created a media environment that allowed for the digital distribution of music.

This book uses 2010 as the end of the project, not to suggest that the digital transformation ended, but rather, to identify 2010 as the year that digital music distribution hit a tipping point. Also significant in 2010 was the Federal Communications Commission's (FCC) adoption of a network neutrality policy, which asserts that Internet service providers (ISPs) cannot charge websites for access to Internet bandwidth (this policy has since been eliminated by the judicial system). Network neutrality is important because there was significant push from the public throughout the 2000s for such a policy. However, throughout the book I point to events that occurred before 1995 and often explain new digital innovations that have happened since 2010.

9. In *Wired Shut* (2007), Tarleton Gillespie argues that the technology of the Internet does not have a predetermined use, but rather the policies that are formed around technology shape how the Internet can be used. While some people claim that the Internet can bring "progress," Gillespie instead argues that technology always fails to achieve social liberation, only entrapping us further. Gillespie explains that while content owners create a debate among the public over piracy, content owners have already implemented policies that secure their position. "At the core of these changes is a fundamental shift in strategy," Gillespie argues, "from regulating the use of technology through law to regulating the design of the technology so as to constrain use" (Gillespie 2007, 6).

10. Throughout this book, I use the term *ideology* to mean the set of ideas that obscure the relations of production in any society. In Marx's terms, ideology obscures the fact that these ideas are the ruling ideas of society. "The ruling ideas are nothing more than the ideal expression of the dominant material relationships, the dominant material relationships grasped as ideas; hence of the relationships which make the one class the ruling one, therefore, the ideas of its dominance" (Marx and Engels 1978, 173). However, I use the term in a slightly broader sense to reflect that ideology is not only a set of ideas created by the dominant class, but also that other classes and cultures can produce ideology in an attempt to negotiate the ruling class's ideology—in this sense, they are counter-ideologies. At the same time, these counter-ideologies are always already a product of the ruling set of ideas.

11. Changes in the relations of production in the music industry have followed similar changes in other industries and throughout society at a moment that is often characterized as the onset of neoliberalism. Neoliberalism, as David Harvey describes, is "a theory of political economic practices that proposes that human well-being can best be advanced by liberating individual entrepreneurial freedoms and skills within an institutional framework characterized by strong private property rights, free markets and free trade" (2005, 2). The thrust of neoliberalism has been to reduce state regulation to create more economic freedom and individual liberty; this ideology asserts that this is the only way to create greater social and economic equality. In reality, Harvey argues, neoliberalism results in "accumulation by dispossession," which for Harvey is a contemporary form of what Marx described as "primitive accumulation" because the capitalism expands through privatization, enclosing intellectual property and bringing latent labor (workers that are not fully part of capitalist production) into capitalism (Harvey 2005). Neoliberals are opposed to state intervention in theory, but industries seek state intervention to create policies that help those industries to generate more profit. Whereas neoliberals claim that intellectual property rights encourage innovation, Harvey explains that intellectual property rights are formed by state intervention on the side of capital to create monopolies. In the recording industry, neoliberal ideology is exemplified in the record labels' request for the state to regulate music with copyright laws.

12. The phrase "relations of production" encompasses both the social aspect of production (i.e., the relationship between capital and labor—record labels and their recording artists) and the means of production (CD pressing plants, marketing departments, recording studios, distributors, etc.). Relations of production also include the recording industry's legal structure, which in the music industry is related directly to copyright law.

13. While there was concern about piracy with tape cassettes, this concern did not reach the level it did with digital music because copies of copies in the tape cassette format lose sound fidelity. This difference is discussed at length in Part II.

14. James Boyle's *The Public Domain: Enclosing the Commons of the Mind* (2008) is an example of an author engaging with economic ideologies about economic incentive to produce intellectual property. In the end, Boyle professes the need for a robust public domain that is not secondary to copyrights, patents, and trademarks. However, his willingness to engage with discussions about incentive to produce intellectual property, by claiming that economic incentive is sometimes necessary, obscures the part of his own argument that this is not always the case.

15. Jonathan Sterne's *MP3: The Meaning of a Format* (2012) traces the history of sound compression through the development of perceptual technics to show the way that ideas about hearing structure what we hear. MP3s are supposed to be adequate representations of music, but digital compression limits the range of sound that we hear.

16. The International Federation of Phonographic Industries is the international trade association of the recording industry—it is the international equivalent of the Recording Industry Association of America.

17. A 360 deal is a recording contract that allows record labels to colonize all aspects of a recording artist's revenue (recording, marketing, touring, appearances, etc.). This is a marked difference from standard recording contracts, which only pertain to the recorded commodity (studio costs, printing albums, etc.). These recording contracts are explained at length below.

1. RECORDING INDUSTRY IN TRANSITION

1. The *Recording Industry in Numbers* reports are published yearly by the IFPI. The data used by regional recording industry trade associations, such as the RIAA or the British Phonographic Industry (BPI), are in these reports. Publication of the reports is narrow in scope, so the recording industry provides a quite different assessment of the state of the industry than can be found in the news media or *Billboard*. Throughout this chapter, I use these reports to analyze the condition of the recording industry based on how it assesses itself; they are my primary source of data.

2. Friedrich Kittler provides a useful conceptualization of transmediation in *Gramophone, Film, Typewriter* (1999).

3. Note that the recording industry consists of the production (recording, engineering, mastering, and printing), marketing/promotion, distribution, and sale (i.e., retail) of *recorded music*.

4. Simon Frith asserts that "hardware companies get involved in software production simply in order to have something on which to demonstrate their equipment" (2006, 233).

5. It is possible to calculate overall music sales prior to 2003 by adding the data provided by Nielsen. However, 2003 is also the year that they began tracking digital music sales; this is not to say that digital music sales were zero in previous years.

6. The production cost of additional songs is negligible for a widely distributed album in relation to the manufacturing and distribution of the CD.

7. Expanding on the work of Raymond Williams (1976), Keith Negus explains that there are three primary senses of mediation that apply to the music industry: (1) "mediation as intermediary action," (2) "mediation as transmission," and (3) "mediation of social relationships" (Negus 1997, 67–69). The first sense relates to the gatekeepers (generally artists and repertoire [A&R] staff, producers, and executives) at record labels who come between the artist and the consumer. In the second sense of the term, mediation has to do with the communication technologies, like CDs and CD players, that come between the artist and the consumer. The third sense focuses on how power as a social relationship structures the way music is heard. In this chapter, it is in the second sense, technology, that mediation is most important in the production of recorded music.

8. The artists and repertoire (A&R) department is the department within a record label that determines the artists that will be signed, the albums that will be recorded, and, ultimately, what the music sounds like.

9. For further discussion of the AHRA, see chapter 2.

10. Jonathan Zittrain argues that this is an effect of the Internet and computer technologies becoming more "closed" (2008).

2. THE EXPANSION OF CONSUMPTION IN THE RECORDING INDUSTRY

1. Hull, Hutchison, and Strasser also begin their explanations of each of three music streams from different copyrights (2011). The way that the copyright generates revenue is epiphenomenal to the ownership and exploitation of the copyright itself.

2. The process through which record labels obtain copyrights from their recording artists through record contracts and how this creates surplus value is explained at length in chapter 3.

3. Michael Jackson's former label was Columbia/Epic Records, a label owned by Sony Music Group.

4. The concept of "public performance" of a work is complicated under current copyright law. Under certain circumstances, there is both a performance royalty and a publishing royalty; however, this is not always the case. Notably, radio broadcasts only give royalties to the author/composer/writer, while the musicians on the recording do not get performance royalties for the public performance of their work.

5. This decline is explainable for a number of reasons, including the recession and a recession-resistant spike in 2009 because of Michael Jackson catalog sales following his death.

6. At this time, revenues and royalties have not been settled entirely on these new formats. There are several court cases going through the judicial system that aim to determine what type of copyright applies to digital distribution. Furthermore, each record label has different contracts negotiated with both its artists and the various services.

7. This supports Miège's ideas about studying different actors within a given cultural industry in order to understand the nuances of that industry. Also see Kraft (2006) for an explanation of how the role of musicians changed in Hollywood following the development and widespread use of mechanical reproduction.

8. However, it is important to note that there was conflict between other arms of the music industry. When the gramophone and phonograph arrived, there was a major conflict between music publishers and record producers over the application of copyright law; this was partially settled in the 1906 copyright law. Furthermore, there were conflicts between music publishers and piano roll manufacturers prior to and during this conflict. I am not painting a conflictless picture, but rather demonstrating that the separate entities of record labels and phonograph manufacturers did not yet exist.

3. COPYRIGHT: A CRITICAL EXPLORATION

1. "Loss leader" refers to a practice by box-store retailers in which they sell music at below wholesale prices—selling the music at a loss—in order to get consumers in the store to buy other goods. This practice in the 1990s led directly to the decline of record stores (Hull, Hutchison, and Strasser 2011).

2. As discussed in chapter 1; see also Adorno (2002b).

3. The Declaration of Independence declared that all "men" have rights that include "life, liberty and the pursuit of happiness." In the American sense, "property" becomes the "pursuit of happiness."

4. Music was a public good because it was both nonrivalrous and part of the public domain before copyrights. This meant that anyone could perform or listen to music without affecting the ability of others to perform or listen to that same music. As part of the public domain, the public holds music authorship and ownership; in other words, music in the public domain is no longer copyrighted. Copyright laws stipulate a limited length of time for which music (or any other copyrighted material) can be copyrighted, and after that period, it goes into the public domain. Everyone can perform and reproduce music that is in the public domain.

5. The specific part of the means of production that musicians and composers sell is their right to own and control the commodities. However, Marx includes the product of a worker's labor as part of the means of production. In this way, the copyrighted material that musicians and composers produce is part of the means of production.

6. The current copyright law is derived from the Copyright Act of 1976 and has been revised several times—some of these revisions are discussed below.

7. In this case, a "public performance" would be a broadcast or playing the music in a public place over loudspeakers.

4. CRITICAL JUNCTURES

1. The DPRA was first introduced in 1993. It took several years and several revisions to pass the law because of tensions among the Recording Industry Association of America, broadcasters, and PROs.

2. This is another point that supports Litman's argument that there is no underlying legal principle behind copyright law (2006).

3. The DMCA was first introduced as a Green Paper in 1994 and a White Paper in September of 1995 by the Lehman Working Group (a working group on the National Information Infrastructure; Patent Commissioner Bruce Lehman heading the group, a Clinton appointee).

4. A group that is primarily composed of law professors and librarians opposed to the DMCA.

5. An advocacy group that supports home audio and videotaping.

6. Even this is highly contestable because of the predominance of U.S./Western/Northern interests. Where this gets complicated is that U.S. imperialism functions through international organizations. The treaties and international laws that are passed by international organizations represent the interests of some (usually corporate) American interests while denying the interests of American citizens.

7. The key point is that digitization works by making copies, and as soon as each circulation of a digital file is defined as a copy, then copyright law comes into play.

8. If someone hacked a computer and stole MP3s, there would be no way to report the files as stolen property. Even more curious with regard to stealing is the music on a stolen MP3 player or computer; if someone has their iPod stolen, that person cannot report the music files contained on the iPod as stolen.

9. The recording industry first introduced the bill that would become the DPRA in 1993, so they were at least aware that an Internet broadcast of music could be the beginning of free music on the Internet.

10. The points made on M4P files here only pertain to the initial way that iTunes restricted its files. Some changes have been made, and Apple continues to change the way that its DRM operates.

11. With the development of iTunes and Amazon.com's "cloud" service, there has been some relaxation on these restrictions; however, the cloud systems have their own restrictions. For one, in order to have access to music in the "cloud," users must pay a fee. Furthermore, access to music in the "cloud" does not overcome DRM on files; it merely allows users to stream music onto their devices. Finally, "cloud" music can only be played in very limited ways, and, for instance, cloud music cannot be used as a ringtone.

12. One group that is a particularly vocal advocate for this position is the Future of Music Coalition.

5. MUSICIAN LABOR

1. For example, Joli Jensen's close reading of public discourse in *Is Art Good for Us?* (2002) shows that the public thinks that not only is art "good" for us, but the reason many people believe this is because they think that artists are in some way autonomous. Furthermore, this idea of autonomy is perpetuated by artistic commentators, as evinced by a blog post entitled "Stevie Wonder: Artistic Autonomy" (Pennington 2011) and an online magazine article entitled "Notes on Alternative Autonomy" (Frock 2011). The post on Stevie Wonder asserts that Wonder was given autonomy by Motown Records at the age of twenty (Pennington 2011) but does not describe *how* Wonder became autonomous. In "Notes on Alternative Autonomy" (2011), Christian Frock assumes that avant-garde art is removed from the control of others.

2. Here Bourdieu is making a distinction between avant-garde art (production for producers) and mass culture (production for mass audiences). Cultural intermediaries play different roles in each realm.

3. Wholesale price − (distribution costs + manufacturing costs + royalties owed) = label profit.

4. Mark Halloran's *Musician's Business & Legal Guide* (2007) does an excellent job walking musicians through this process.

5. I am using the term *wave* here as Vandana Shiva (2008) uses the term to signify the cumulative effect of developing ideas that do not go away but are rather built upon earlier conceptions.

6. Marx describes "necessary labor" as the "part of [the worker's] day's labor in which he produces the value of his labor-power. . . . I call the portion of the working day during which this reproduction takes place necessary labor-time, and the labor expended during that time necessary labor; necessary for the work, because independent of the particular social form of his labor; necessary for capital and the capitalist world, because the continued existence of the worker is the basis of that world" (Marx 1992, 324–25).

7. "The direct form of the circulation of commodities is C—M—C, the transformation of commodities into money and the re-conversion of money into commodities: selling in order to buy" (Marx 1992, 247).

8. Note that this only refers to recorded music. In the broader music industry, surplus value is not confined to recordings or record labels.

9. While I have described this as a relation between capital and labor, the record labels design the relationship to appear as a relationship between capital and capital. Artists are convinced to act as independent firms but are required to take a loan from a more powerful firm to do so. This loan allows the "artist firm" to pay itself wages (i.e., to reproduce its labor) in the hopes of "hitting the big time." In exchange for this loan, however, the artist must surrender property rights to the music itself. The only revenue stream that remains is royalties on album sales. It is upon this single stream that the artist firm is relying to realize surplus value from its own labor.

10. The observations in this paragraph are based on my personal observations as a gigging musician and discussions with other musicians. While live performance is the means by which most musicians earn money in the music industry, few music business writers spend much time discussing the process through which musicians go to contract gigs. There are some useful websites, such as http://www.music-law.com/playlive.html, that help musicians negotiate this process, but there is little academic work done on the relationship between musicians and venues.

6. VICTIMS, MUSICIANS, AND METALLICA

1. A 360 deal is a recording contract that allows record labels to colonize all aspects of a recording artist's revenue (recording, marketing, touring, appearances, etc.). This is a marked

difference from standard recording contracts, which only pertain to the recorded commodity (studio costs, printing albums, etc.). These recording contracts are explained at length below.

2. Simon argues that "governing through crime" is strategic for politicians running for office. What this means is that people running for office can always take a "tough on crime" stance and receive support from various victims groups. Once a candidate takes this position, his or her opponent must take this position as well because no one can be seen as soft on crime and win an election.

3. McGuinn and Barry were included in the hearing to support Napster.

4. At the same time, it is important to recognize that Metallica does have more autonomy than most recording artists because of their position as one of the top-selling bands in history. As recording artists gain popularity, they can often renegotiate contracts after several albums to get higher royalties and more autonomy from the label. In the case of Metallica, this actually resulted in the band suing its label, Elektra, and settling out of court, but the details of the settlement have not been released because of a nondisclosure agreement (Fricke 1995).

7. DISTRIBUTION THEN AND NOW

1. Parts of this chapter have appeared previously in "Disciplining the Consumer: File-Sharers under the Watchful Eye of the Music Industry" (Arditi 2011).

2. There is a long history of debate over the relationship between "value" and "price"—a debate called the "transformation problem" (Oakley 1976). Most neoclassical economists dispute any simple equation between the amount of labor time devoted to production and price. However, all other things being equal, the more labor power devoted to the commodity, the higher the production cost, and therefore the higher the price.

3. This is to say nothing of the role of record label marketing departments, which give a significant advantage to major label record artists over independent artists.

4. Major record labels still report album shipments instead of actual album sales. The RIAA's data still reflect shipment data instead of sales data.

5. See http://www.riaa.com under the "Piracy" tab.

6. In *Sony Corp. of America v. Universal City Studios, Inc.*, 464 U.S. 417 (also known as the Betamax Case), the Supreme Court ruled that videotaping an entire television show was legal under fair use because it allowed for "time shifting"—watching a television show at a different time than it originally aired. Time shifting, the court ruled, is not equivalent to the commercial reproduction of copyrighted material.

7. A cloud service is like a digital locker. It allows users to access their files on a remote digital storage unit. Today, Google, Apple, and Amazon have popular cloud services for users to upload their files to and access remotely.

8. Foucault's term *governmentality* names "a concept whose whole rationale was to grasp the birth and characteristics of a whole variety of ways of problematizing and acting on individual and collective conduct in the name of certain objectives which do not have the State as their origin or point of reference" (Rabinow and Rose 2006).

9. I am focusing on P2P programs here because the surveillance of these programs is directly related to the lawsuits against individuals even though surveillance on the Internet now incorporates much more than the behavior of people on P2P programs.

10. Interestingly, when I began work on this project, I referenced a website called Amie Street, launched in 2006, that made music available by unsigned artists. However, Amie Street was bought by Amazon.com in 2010, and subsequently, Amazon.com eliminated this service.

11. Napster merged with Rhapsody, another music subscription service, in 2011. While this occurred after the period of this project, it is worth noting because music subscription services remain quite fluid as different services compete and merge. The issue with these subscription services merging and closing so frequently is that users who subscribe lose their music libraries when their service closes.

12. Penn State University changed its contract from Napster to Ruckus in 2007. Subsequently, Ruckus was purchased by a joint venture between Universal Music Group and Sony Music but was later closed in 2008.

13. Note that there are ways to play iTunes as ringtones. DRM-free iTunes files can be converted to a ringtone file using music production programs such as Garage Band. However, in order for people to do this, they need the technical knowledge to convert the files, the program, and the knowledge that this process is possible.

8. WATCHING MUSIC CONSUMPTION

1. Parts of this chapter have appeared previously in "Disciplining the Consumer: File-Sharers under the Watchful Eye of the Music Industry" (Arditi 2011).

2. According to a study conducted by the Pew Internet and American Life Project between 2006 and 2009, Internet users have become increasingly aware that information about them is available online (Madden and Smith 2010). Whenever Google or Facebook change their privacy settings, newspapers publish articles about these changes, and at least within my circle of "friends," these articles are posted on Facebook. There is a thorough discussion of Internet users' perception of the RIAA's surveillance below.

3. The iTunes Store opened in April 2003, in June the RIAA began to warn file sharers that lawsuits would commence, and the lawsuits against P2P users began in September of 2003.

4. Smythe uses the category "Consciousness Industry" instead of "Culture Industry."

5. A song that is in the iTunes format can be converted to WAV format and burned onto a disc, but the digital song owner must be careful not to copy the file in the original format because this will count as one of the copies allowed by DRM. There are other ways to get around DRM, but they are time intensive and often require an advanced level of computer skill.

6. These are data that are stored in the file at the moment of purchase and can include information such as the downloader's name and email address.

7. Despite the fact that iTunes and Amazon use DRM-free music files, DRM is still a critical part of the recording industry's new business model. Files from iTunes and Amazon are "DRM-free" in name only (as discussed above), and music from other sources still contain DRM.

CONCLUSION

1. The RIAA dates the beginning of the decline to 2000, although Nielsen SoundScan dates it to 2001.

Bibliography

"500 Are Out in First Wave of US Layoffs." 1999. *Billboard*, January 30.

Adegoke, Yinka. 2014. "Cashing In on Cover Versions." *Billboard*, June 27.

Adorno, Theodor W. 2002a. "The Curves of the Needle." In *Essays on Music*, edited by Theodor W. Adorno, Richard D. Leppert, and Susan H. Gillespie, 271–76. Berkeley: University of California Press.

———. 2002b. "The Form of the Phonograph Record." In *Essays on Music*, edited by Theodor W. Adorno, Richard D. Leppert, and Susan H. Gillespie, 277–80. Berkeley: University of California Press.

Ahrens, Frank. 2002. "Stars Come Out against Net Music Piracy in New Ads." *Washington Post*, September 26.

Alderman, John. 2002. *Sonic Boom*. New York: Basic Books.

Amin, Ash, ed. 1995. *Post-Fordism: A Reader*. Cambridge, MA: Wiley-Blackwell.

Andersen, Birgitte, and Marion Frenz. 2010. "Don't Blame the P2P File-Sharers: The Impact of Free Music Downloads on the Purchase of Music CDs in Canada." *Journal of Evolutionary Economics* 20 (5): 715–40.

Andrejevic, Mark. 2007. *iSpy: Surveillance and Power in the Interactive Era*. CultureAmerica. Lawrence: University Press of Kansas.

———. 2011. "Exploitation in the Data Mine." In *Internet and Surveillance: The Challenges of Web 2.0 and Social Media*, edited by Christian Fuchs, Kees Boersma, Anders Albrechtslund, and Marisol Sandoval, 71–88. New York: Routledge.

Arditi, David. 2007. *Criminalizing Independent Music: The Recording Industry Association of America's Advancement of Dominant Ideology*. Saarbrücken, Germany: VDM Verlag.

———. 2011. "Disciplining the Consumer: File-Sharers under the Watchful Eye of the Music Industry." In *Internet and Surveillance: The Challenges of Web 2.0 and Social Media*, edited by Christian Fuchs, Kees Boersma, Anders Albrechtslund, and Marisol Sandoval, 170–86. New York: Routledge.

———. 2014. "iTunes: Breaking Barriers and Building Walls." *Popular Music and Society* 37 (4): 408–24.

Associated Press. "Amazon Buys Online Music Retailer Amie Street." 2010. September 8.

Attali, Jacques. 1985. *Noise: The Political Economy of Music*. Theory and History of Literature. Minneapolis: University of Minnesota Press.

Atwood, Brett. 1995. "Advanced Broadcast System to Transmit via Computers." *Billboard*, March 11.

Ault, Susanne. 2002. "Def Jam, EA Create Hip-Hop Video Line." *Billboard*, September 7.

Avalon, Moses. 1998. *Confessions of a Record Producer: How to Survive the Scams and Shams of the Music Business*. San Francisco: Miller Freeman Books.

Baker, C. Edwin. 2007. *Media Concentration and Democracy: Why Ownership Matters*. Communication, Society, and Politics. New York: Cambridge University Press.

Bate, Tom. 1998. "Diamond Multimedia Systems Is Caught in Legal Quagmire." *San Francisco Chronicle*, October 13.

Benkler, Yochai. 2006. *The Wealth of Networks: How Social Production Transforms Markets and Freedom*. New Haven, CT: Yale University Press.

Bhattacharjee, Sudip, Ram Gopal, Kaveepan Lertwachara, and James Marsden. 2006. "Impact of Legal Threats on Online Music Sharing Activity: An Analysis of Music Industry Legal Actions." *Journal of Law and Economics* 49 (April): 91–114.

"Bits & Briefs." 2010. *Billboard*, November 20.

Block, Debbie. 1995. "Curbing Piracy in a Digitally Perfect World." *Tape-Disc Business*, June 1.

Boehlert, Eric. 1994. "Radio Biz Finds New Way to Network." *Billboard*, March 26.

Boliek, Brooks. 1997. "RIAA Seeks 40% Digital Royalty." *Hollywood Reporter*, December 12.

Bourdieu, Pierre. 1984. *Distinction: A Social Critique of the Judgement of Taste*. Cambridge, MA: Harvard University Press.

———. 1993. *The Field of Cultural Production: Essays on Art and Literature*. Edited by Randal Johnson. New York: Columbia University Press.

Boyle, James. 2008. *The Public Domain: Enclosing the Commons of the Mind*. New Haven, CT: Yale University Press.

Branch, Al. 2009. "The Deals of the Future." *Billboard*, November 21.

Bruno, Antony. 2008a. "Showdown Looming." *Billboard*, September 6.

———. 2008b. "Battle of the Brands: MTV vs. Activision—Game On!" *Billboard*, November 8.

———. 2009. "A New Game Plan." *Billboard*, January 10.

———. 2010a. "What Can Apps Do for You?" *Billboard*, April 17.

———. 2010b. "Beatles Catalog Finally Coming to iTunes, Apple Announces." *Billboard*, November 16.

Bruno, Antony, and Susan Butler. 2008. "Auto Focus: As 'Grand Theft Auto' Rewrites Gaming History, the Music Biz Gains Big." *Billboard*, May 3.

Burkart, Patrick. 2010. *Music and Cyberliberties*. Middletown, CT: Wesleyan.

Burkart, Patrick, and Tom McCourt. 2006. *Digital Music Wars: Ownership and Control of the Celestial Jukebox*. Lanham, MD: Rowman & Littlefield.

Burlingame, Jon. 1997. *For the Record: The Struggle and Ultimate Political Rise of American Recording Musicians within Their Labor Movement*. Hollywood, CA: RMA Recording Musicians Association.

Butler, Susan. 2004. "Ashcroft Praises Biz's Piracy Fight." *Billboard*, December 25.

———. 2007. "Uncovering Cover Versions: '80s Rockers the Romantics Throw Crimp into Gaming Plans." *Billboard* 119 (50): 13–13.

Caldwell, John. 2004. "Convergence Television: Aggregating the Form and Repurposing Content in the Culture of Conglomeration." In *Television after TV: Essays on a Medium in Transition*, edited by Lynn Spigel and Jan Olsson. Durham, NC: Duke University Press.

Chapple, Steve, and Reebee Garofalo. 1977. *Rock "N" Roll Is Here to Pay: The History and Politics of the Music Industry*. Chicago: Nelson-Hall.

Christman, Ed. 1998. "Merging Two Distributors into One." *Billboard*, December 19.

———. 2002. "Handleman and Anderson Merchandisers Are Hustling to Keep the Mass Merchants Stocked with Hit Music." *Billboard*, August 31.

———. 2004. "Atlantic Slims Down." *Billboard*, April 10.

———. 2010. "Digital Divide: Apple Solidifies Its Lead among U.S. Music Accounts, as Mobile Merchants Fade." *Billboard*, May 22.

———. 2014a. "The Digital Decline: What's behind the First Downturn of the Itunes Era? And Can Streaming Save the Day?" *Billboard*, January 18.

———. 2014b. "Newer Rules." *Billboard*, January 18.

Cloonan, Martin, and Reebee Garofalo. 2003. *Policing Pop*. Sound Matters. Philadelphia: Temple University Press.

Connolly, Marie, and Alan B. Krueger. 2005. *Rockonomics: The Economics of Popular Music.* SSRN Scholarly Paper 711924. Rochester, NY: Social Science Research Network.

Coombe, Rosemary J. 1998. *The Cultural Life of Intellectual Properties: Authorship, Appropriation, and the Law.* Post-Contemporary Interventions. Durham, NC: Duke University Press.

Csathy, Peter. 2014. "Why This Venture Capitalist Is Optimistic about the Music Business." *Billboard*, February 8.

Dahl, Robert A. 2005. *Who Governs? Democracy and Power in an American City.* 2nd ed. New Haven, CT: Yale University Press.

David, Matthew. 2010. *Peer to Peer and the Music Industry: The Criminalization of Sharing.* Los Angeles: Sage Publications.

Debord, Guy. 1994. *The Society of the Spectacle.* New York: Zone Books.

Donnelly, Bob. 2010. "Buyer Beware." *Billboard*, March 27.

Elberse, Anita. 2013. *Blockbusters: Hit-Making, Risk-Taking, and the Big Business of Entertainment.* New York: Henry Holt.

Florida, Richard L. 2004. *The Rise of the Creative Class: And How It's Transforming Work, Leisure, Community and Everyday Life.* New York: Basic Books.

Flynn, Laurie. 2004. "The Cellphone's Next Makeover: Affordable Jukebox on the Move." *New York Times*, August 2. http://www.nytimes.com/2004/08/02/business/technology-the-cellphone-s-next-makeover-affordable-jukebox-on-the-move.html.

Forest, Greg. 2008. *The Music Business Contract Library.* New York: Hal Leonard.

Foucault, Michel. 1977. *Discipline and Punish: The Birth of the Prison.* New York: Pantheon.

Fricke, David. 1995. "Married to Metal." *Rolling Stone*, May 18.

———. 2012. "The Man with the Magic Ears." *Rolling Stone*, April 12.

Frith, Simon. 2006. "The Industrialization of Music." In *The Popular Music Studies Reader*, edited by Andy Bennett, Barry Shank, and Jason Toynbee, 231–38. New York: Routledge.

Frock, Christian. 2011. "Notes on Alternative Autonomy." *Art Practical*, March 24. http://www.artpractical.com/feature/notes_on_alternative_autonomy/.

Fuchs, Christian. 2008. *Internet and Society: Social Theory in the Information Age.* New York: Routledge.

———. 2010. "Labor in Informational Capitalism and on the Internet." *Information Society* 26 (3): 179–96.

Fung, Brian. 2014. "Federal Appeals Court Strikes down Net Neutrality Rules." *Washington Post*, January 14. http://www.washingtonpost.com/blogs/the-switch/wp/2014/01/14/d-c-circuit-court-strikes-down-net-neutrality-rules/.

Garnham, Nicholas, and Fred Inglis. 1990. *Capitalism and Communication: Global Culture and the Economics of Information.* Newbury Park, CA: Sage Publications.

Garrity, Brian. 2007. "Adding up iTunes Plus." *Billboard*, June 23.

Garrity, Brian, and Ed Christman. 2003. "RIAA Figures Show Continuing Decline." *Billboard*, September 13.

Gillespie, Tarleton. 2007. *Wired Shut: Copyright and the Shape of Digital Culture.* Cambridge, MA: MIT Press.

Goodwin, Andrew. 1992. "Rationalization and Democratization in the New Technologies of Popular Music." In *Popular Music and Communication*, edited by James Lull, 75–100. Newbury Park, CA: Sage Publications.

Gordon, Steve. 2008. *The Future of the Music Business: How to Succeed with the New Digital Technologies.* 2nd ed. Milwaukee: Hal Leonard Music Pro Guides.

Graham, Lee. 2009. *Continued Sales Declines in 2008, but Music Listening and Digital Downloads Increase.* Marketing Report. NPD Digital Music Study. Port Washington, NY: NPD Group.

Gramsci, Antonio. 1971. *Selections from the Prison Notebooks of Antonio Gramsci.* Edited by Quintin Hoare and Geoffrey Nowell-Smith. London: Lawrence & Wishart.

Greeley, Brendan. 2011. "Daniel Ek's Spotify: Music's Last Best Hope." *Business Week*, July 13. http://www.businessweek.com/magazine/daniel-eks-spotify-musics-last-best-hope-07142011.html.

Green, Stuart P. 2012. "'Theft' Law in the 21st Century." *New York Times*, March 28. http://
www.nytimes.com/2012/03/29/opinion/theft-law-in-the-21st-century.html.

Greene, Andy. 2012. "The Who Sell Out: Townshend Gets Millions for Back Catalog." *Rolling
Stone*, March 1.

Hall, Tia. 2002. "Music Piracy and the Audio Home Recording Act." *Duke Law & Technology
Review* 1 (1): 1–8.

Halloran, Mark. 2007. *Musician's Business & Legal Guide*. 4th ed. Upper Saddle River, NJ:
Prentice Hall.

Harding, Cortney. 2010. "Maximum Exposure: DIY." *Billboard*, October 2.

Harrison, Anthony Kwame. 2006. "'Cheaper Than a CD, Plus We Really Mean It': Bay Area
Underground Hip Hop Tapes as Subcultural Artifacts." *Popular Music* 25 (2): 283–301.

Harvey, David. 2005. *A Brief History of Neoliberalism*. New York: Oxford University Press.

———. 2010. *A Companion to Marx's Capital*. Brooklyn, NY: Verso.

Hatch, Orin. 2000. *Music on the Internet: Is There an Upside to Downloading?* Washington,
DC: U.S. Government Printing Office.

Herman, Edward S., and Noam Chomsky. 2002. *Manufacturing Consent: The Political Economy of the Mass Media*. New York: Pantheon.

Hesmondhalgh, David. 2006. "The British Dance Music Industry: A Case Study of Independent Cultural Production." In *The Popular Music Studies Reader*, edited by Andy Bennett,
Barry Shank, and Jason Toynbee, 247–52. New York: Routledge.

———. 2007. *The Cultural Industries*. 2nd ed. Los Angeles: SAGE.

Hiatt, Brian. 2013. "2013: The Future Has Arrived." *Rolling Stone*, December 19.

Hoglund, Don. 2011. "RIAA Equalization Curve for Phonograph Records." *Granite Audio*.
http://www.graniteaudio.com/phono/page5.html.

Holland, Bill. 1995. "Performance Right Act Becomes Law." *Billboard*, November 11.

Howe, Jeff. 2003. "BigChampagne Is Watching You." *Wired.com*.http://www.wired.com/
wired/archive/11.10/fileshare.html.

Hull, Geoffrey P. 2004. *The Recording Industry*. 2nd ed. New York: Routledge.

Hull, Geoffrey P., Thomas W. Hutchison, and Richard Strasser. 2011. *The Music Business and
Recording Industry: Delivering Music in the 21st Century*. 3rd ed. New York: Routledge.

IFPI. 1996. *Recording Industry in Numbers*. International Federation of the Phonographic
Industry.

———. 2001. *Recording Industry in Numbers*. International Federation of the Phonographic
Industry.

———. 2002. *Recording Industry in Numbers*. International Federation of the Phonographic
Industry.

———. 2003. *Recording Industry in Numbers*. International Federation of the Phonographic
Industry.

———. 2004. *Recording Industry in Numbers*. International Federation of the Phonographic
Industry.

———. 2005. *Recording Industry in Numbers*. International Federation of the Phonographic
Industry.

———. 2007. *Recording Industry in Numbers*. International Federation of the Phonographic
Industry.

———. 2008. *Recording Industry in Numbers*. International Federation of the Phonographic
Industry.

———. 2009. *Recording Industry in Numbers*. International Federation of the Phonographic
Industry.

———. 2010. *Recording Industry in Numbers*. International Federation of the Phonographic
Industry.

———. 2011. *Recording Industry in Numbers*. International Federation of the Phonographic
Industry.

———. 2012. *Recording Industry in Numbers*. International Federation of the Phonographic
Industry.

Jeffrey, Don. 1997. "Downloading Songs Subject of RIAA Suit." *Billboard*, June 21.

Jenkins, Henry. 2006. *Convergence Culture: Where Old and New Media Collide*. New York: New York University Press.

Jensen, Joli. 2002. *Is Art Good for Us? Beliefs about High Culture in American Life*. Lanham, MD: Rowman & Littlefield.

Jones, Steve. 2002. "Music That Moves: Popular Music, Distribution and Network Technologies." *Cultural Studies* 16 (2): 213–32.

Kedrosky, Paul. 2000. "Online or Offline, Theft Is Theft." *National Post*, August 1.

Kittler, Friedrich A. 1999. *Gramophone, Film, Typewriter*. Stanford, CA: Stanford University Press.

Klein, Naomi. 2002. *No Logo: No Space, No Choice, No Jobs*. New York: Picador.

Knopper, Steve. 2007. "Reinventing Record Deals." *Rolling Stone*, November 29.

———. 2009. *Appetite for Self-Destruction: The Spectacular Crash of the Record Industry in the Digital Age*. New York: Free Press.

———. 2011. "The New Economics of the Music Industry: How Artists Really Make Money in the Cloud—or Don't." *Rolling Stone*, October 25. http://www.rollingstone.com/music/news/the-new-economics-of-the-music-industry-20111025.

———. 2012. "Is the CD Era Finally Over? Labels Take Tough Look at Format as Sales, Profits Continue to Fall." *Billboard*, March 1.

———. 2013. "Jay-Z Stumbles." *Rolling Stone*, August 1.

Kraft, James P. 2006. "Musicians in Hollywood: Work and Technological Change in Entertainment Industries, 1926–1940." In *The Popular Music Studies Reader*, edited by Andy Bennett, Barry Shank, and Jason Toynbee, 239–45. New York: Routledge.

Krasilovsky, M. William, Sidney Shemel, John M Gross, and Jonathan Feinstein. 2007. *This Business of Music*. 10th ed. New York: Billboard Books.

Langenderfer, Jeff, and Don Lloyd Cook. 2001. "Copyright Policies and Issues Raised by A& M Records v. Napster: 'The Shot Heard "Round the World" or "Not with a Bang but a Whimper?"'" *Journal of Public Policy & Marketing* 20 (2): 280–88.

Lawrence, Ava. 2004. "Market Research in the Internet Age: How Record Companies Will Profit from Illegal File-Sharing." *Journal of the Music & Entertainment Industry Educators Association* 4 (1): 29–40.

Lessig, Lawrence. 2004. *Free Culture: The Nature and Future of Creativity*. New York: Penguin Press.

———. 2006. *Code: Version 2.0*. 2nd ed. New York: Basic Books.

Lettice, John. 2009. "Apple iTunes Store Goes '100% DRM-Free'—Allegedly." *Register*. http://www.theregister.co.uk/2009/01/06/macworld_itunes/.

Lieb, Kristin. 1994. "Labels Look to Lead Biz's Technology Revolution." *Billboard*, April 16.

Litman, Jessica. 2003. "Sharing and Stealing." *Social Science Research Network*. http://papers.ssrn.com/sol3/papers.cfm?abstract_id=472141.

———. 2006. *Digital Copyright*. Amherst, NY: Prometheus Books.

Lyon, David. 2007. *Surveillance Studies: An Overview*. Malden, MA: Polity.

Madden, Mary. 2009. "The State of Music Online: Ten Years after Napster." Pew Internet and American Life Project. Pew Research Center. http://www.pewinternet.org/2009/06/15/the-state-of-music-online-ten-years-after-napster/.

Madden, Mary, and Amanda Lenhart. 2004. "Sharp Decline in Music File Swappers." Pew Internet and American Life Project. Pew Research Center. http://www.pewinternet.org/2004/01/04/sharp-decline-in-music-file-swappers/.

Madden, Mary, and Lee Rainie. 2005. "Music and Video Downloading." Pew Internet and American Life Project. Pew Research Center. http://www.pewinternet.org/files/old-media/Files/Reports/2005/PIP_Filesharing_March05.pdf.pdf.

Madden, Mary, and Aaron Smith. 2010. *How People Monitor Their Identity and Search for Others Online*. Pew Internet and American Life Project. Pew Research Center. http://www.pewinternet.org/2010/05/26/reputation-management-and-social-media/.

Marcuse, Herbert. 1991. *One-Dimensional Man: Studies in the Ideology of Advanced Industrial Society*. Boston: Beacon Press.

Marshall, Lee. 2005. *Bootlegging: Romanticism and Copyright in the Music Industry*. London: Sage Publications.

Marx, Karl. 1992. *Capital, Volume 1: A Critique of Political Economy.* New York: Penguin Classics.

———. 2000. "Wage-Labour and Capital." In *Selected Writings,* 2nd ed., edited by David McLellan, 273–94. New York: Oxford University Press.

Marx, Karl, and Friedrich Engels. 1978. "The German Ideology." In *The Marx-Engels Reader,* edited by Robert C. Tucker, 146–202. New York: W. W. Norton.

McChesney, Robert Waterman. 2004. *The Problem of the Media: U.S. Communication Politics in the Twenty-First Century.* New York: Monthly Review Press.

———. 2007a. *Communication Revolution: Critical Junctures and the Future of Media.* New York: New Press.

———. 2007b. "Freedom of the Press for Whom? The Question to Be Answered in Our Critical Juncture." *Hofstra Law Review* 35 (3): 1433–54.

McChesney, Robert Waterman, and John Foster. 2011. "The Internet's Unholy Marriage to Capitalism." *Monthly Review,* March. http://monthlyreview.org/2011/03/01/the-internets-unholy-marriage-to-capitalism.

McCollum, Brian. 2007. "Romantics Band Sues 'Guitar Hero' Publisher." *USA Today,* November 22. http://www.usatoday.com/life/music/2007-11-22-romanticssue_N.htm.

McDaniels, Robb. 2014. "Please Adjust Your Bet." *Billboard,* January 25.

McLeod, Kembrew. 2005. "MP3s Are Killing Home Taping: The Rise of Internet Distribution and Its Challenge to the Major Label Music Monopoly." *Popular Music and Society* 28 (4): 521–31.

Miller, Jim. 1999. *Flowers in the Dustbin: The Rise of Rock and Roll, 1947–1977.* New York: Simon & Schuster.

Miller, Roger Leroy, Gaylord A. Jentz, Frank B. Cross, Kenneth W. Clarkson. 2000. *West's Business Law: Text and Cases—Legal, Ethical, Regulatory, International, and E-Commerce Environment.* Edited by Kenneth W. Clarkson. 8th ed. Nashville: South-Western.

Mosco, Vincent. 2009. *The Political Economy of Communication.* 2nd ed. Los Angeles: Sage Publications.

Negus, Keith. 1997. *Popular Music in Theory: An Introduction.* Hanover, NH: University Press of New England.

———. 1999. *Music Genres and Corporate Cultures.* New York: Routledge.

Newman, Melinda. 2005. "SUM, Columbia Bolster A&R Efforts." *Billboard,* August 27.

Newman, Simon. 1997. "Intellectual Property Law—Rights, Freedoms and Phonograms: Moral Rights and Adaptation Rights in Music and Other Copyright Works." *Computer Law & Security Report* 13 (1): pp. 22–28.

Nielsen. 1997. *SoundScan 1996 Year-End Music Industry Report.*

Nielsen. 1999. *SoundScan 1998 Year-End Music Industry Report.*

Nielsen. 2001. *SoundScan 2000 Year-End Music Industry Report.*

Nielsen. 2011. *SoundScan 2010 Year-End Music Industry Report.*

Oakley, Allen. 1976. "Two Notes on Marx and the 'Transformation Problem.'" *Economica,* n.s., 43 (172): 411–17.

Oberholzer-Gee, Felix, and Koleman Strumpf. 2007. "The Effect of File Sharing on Record Sales: An Empirical Analysis." *Journal of Political Economy* 115 (1): 1–42.

Ogg, Erica. 2010. "The Beatles Come to iTunes at Last." *CNET News,* November 16. http://news.cnet.com/8301-31021_3-20022922-260.html.

Park, David J. 2007. *Conglomerate Rock: The Music Industry's Quest to Divide Music and Conquer Wallets.* Lanham, MD: Lexington Books.

Patel, Marilyn Hall. 2000. *A&M Records, Inc. v. Napster, Inc.* Judge Patel's Ruling. United States District Court for the Northern District of California.

Patry, William. 2009. *Moral Panics and the Copyright Wars.* New York: Oxford University Press.

Paul Edmond Dowling v. United States. 1985. United States Supreme Court.

Pegoraro, Rob. 2002. "Copyright Concerns Lead the Year's Big Fusses and Flaps." *Washington Post,* December 29.

Pennington, Paul. 2011. "Stevie Wonder: Artistic Autonomy." *Revivalist.* http://revivalist.okayplayer.com/2011/09/19/stevie-wonder-artistic-autonomy/.

Peoples, Glenn. 2010. "Packaged Goods." *Billboard*, October 23.

Pfanner, Eric. 2010. "Digital Music Gains, but Can't Offset Low CD Sales; Music Industry Group Blames Digital Piracy for 5-Year Global Drop." *International Herald Tribune*, January 22.

Pham, Alex. 2013. "Stop Playing Games." *Billboard*, November 9.

———. 2014. "Loud and Clearer." *Billboard*, January 25.

Pham, Alex, Andrew Hampp, and Ed Christman. 2013. "New Tools." *Billboard*, July 20.

Poster, Mark. 1990. *The Mode of Information: Poststructuralism and Social Context*. Chicago: University of Chicago Press.

Rabinow, Paul, and Nikolas Rose. 2006. "Biopower Today." *BioSocieties* 1 (2): 195–217.

Rainie, Lee, Graham Mudd, Mary Madden, and Dan Hess. 2004. "14% of Internet Users Say They No Longer Download Music Files." Pew Internet and American Life Project. Pew Research Center. http://www.pewinternet.org/2004/04/25/14-of-internet-users-say-they-no-longer-download-music-files/.

Ransby, Barbara. 2003. *Ella Baker and the Black Freedom Movement: A Radical Democratic Vision*. Chapel Hill: University of North Carolina Press.

Reardon, Marguerite, and Greg Sandoval. 2009. "Verizon Tests Sending Copyright Notices." *CNET*. http://news.cnet.com/8301-1023_3-10396787-93.html?tag=newsEditorsPicksArea.0.

Reece, Doug, and Don Jeffrey. 1998. "Labels Striving for Security in the Digital Future: Biz Teams Up to Create Online Distribution Standard." *Billboard*, December 26.

Rheingold, Howard. 2000. *The Virtual Community: Homesteading on the Electronic Frontier*. Rev. ed. Cambridge, MA: MIT Press.

RIAA. 2012. "Scope of the Problem." *RIAA.com*. http://riaa.com/physicalpiracy.php?content_selector=piracy-online-scope-of-the-problem.

Rosen, Hilary. 2000. "It's Outright Theft." *San Jose Mercury News*, July 24.

Rosenstiel, Tom. 2009. *The State of the News Media 2009*. Project for Excellence in Journalism. Pew Research Center. http://stateofthemedia.org/2009/.

Rosenstiel, Tom, and Amy Mitchell. 2011. *The State of the News Media 2011*. Project for Excellence in Journalism. Pew Research Center. http://stateofthemedia.org/overview-2011/.

Ross, Andrew. 2003. *No-Collar: The Humane Workplace and Its Hidden Costs*. New York: Basic Books.

Rothenbuhler, Eric W., and Tom McCourt. 2006. "Commercial Radio and Popular Music: Processes of Selection and Factors of Influence." In *The Popular Music Studies Reader*, edited by Andy Bennett, Barry Shank, and Jason Toynbee, 309–16. London: Routledge.

Sandoval, Greg. 2007. "Mother Protects YouTube Clip by Suing Prince." *CNET*, October 30. http://news.cnet.com/8301-10784_3-9807555-7.html.

———. 2011. "Bye-Bye, Physical Media? Sony Closes CD Plant." *CNET News*. http://www.cnet.com/news/bye-bye-physical-media-sony-closes-cd-plant/.

Saxe, Frank. 2001a. "NAB Files New Suit over Net Streaming." *Billboard*, February 10.

———. 2001b. "Radio, Record Labels Chafe over Streaming." *Billboard*, May 26.

Schiller, Dan. 1996. *Theorizing Communication: A History*. New York: Oxford University Press.

———. 2000. *Digital Capitalism: Networking the Global Market System*. Cambridge, MA: MIT Press.

Scott, Allen John. 2000. *The Cultural Economy of Cities: Essays on the Geography of Image-Producing Industries*. London: Sage Publications.

Segal, David. 2000. "Pop Notes." *Washington Post*, April 19.

Serjeant, Jill. 2010. "Beatles Sell over 2 Million in First Week on iTunes." *Reuters*, November 24.

Shiva, Vandana. 2008. "Ecological Balance in an Age of Globalisation." In *The Globalization Reader*, edited by Frank J. Lechner and John Boli, 465–73. Malden, MA: Blackwell.

Simon, Jonathan. 2007. *Governing through Crime: How the War on Crime Transformed American Democracy and Created a Culture of Fear*. New York: Oxford University Press.

Slichter, Jacob. 2004. *So You Wanna Be a Rock & Roll Star: How I Machine-Gunned a Roomful of Record Executives and Other True Tales from a Drummer's Life*. New York: Broadway Books.

Smith, Paul. 1997a. *Millennial Dreams: Contemporary Culture and Capital in the North*. Haymarket Series. New York: Verso.

———. 1997b. "Tommy Hilfiger in the Age of Mass Customization." In *No Sweat: Fashion, Free Trade, and the Rights of Garment Workers*, edited by Andrew Ross, 249–62. London: Verso.

Smythe, Dallas Walker. 1981. "On the Audience Commodity and Its Work." In *Dependency Road: Communications, Capitalism, Consciousness, and Canada*, edited by Dallas Walker Smythe, 22–51. Norwood, NJ: Ablex.

Starrett, Robert. 1999. "RIAA Loses Bid for Injunction to Stop Sale of Diamond Multimedia RIO MP3 Player, Appeal Pending." *EMedia Professional*, January.

Sterne, Jonathan. 2012. *MP3: The Meaning of a Format*. Durham, NC: Duke University Press Books.

Subramanian, Courtney. 2013. "Beyoncé Breaks iTunes Record with New Album." *Time.com*, December 17.

Thigpen, David, and Jenny Eliscu. 2000. "Metallica Slams Napster." *Rolling Stone*, May 25.

Thompson, Kristin. 2012. *Are Musicians Benefiting from Music Tech?* Artist Revenue Streams. Future of Music Coalition.

"Top Managers Mull the State of the 'UniGram' Union." 1998. *Billboard*, October 10.

Traiman, Steve. 2003. "Licensed to Sell: Artists' Tracks, Likenesses & Alter-Egos." *Billboard*, June 14.

Ulaby, Neda. 2003. "Death of the 'Concept' Album?" *All Things Considered*. National Public Radio. http://www.npr.org/templates/story/story.php?storyId=1554633.

Van Buskirk, Eliot. 2009. "Inside BigChampagne's Music Panopticon." *Wired.com*. http://www.wired.com/epicenter/2009/08/inside-bigchampagnes-music-panopticon/.

Washington Post. 2011. "Access Industries to Buy Warner Music." May 7, Suburban edition, section A.

Watkins, S. Craig. 2005. *Hip Hop Matters: Politics, Pop Culture, and the Struggle for the Soul of a Movement*. Boston: Beacon Press.

Williams, Raymond. 1976. *Keywords: A Vocabulary of Culture and Society*. London: Fontana.

Yardley, Jonathan. 2000. "The Napster Generation." *Washington Post*, May 8.

"Year-End Industry Shipment and Revenue Statistics." 2013. Database. *RIAA Shipment Database*. http://www.riaa.com/chartindex.php.

Zittrain, Jonathan. 2008. *The Future of the Internet and How to Stop It*. New Haven, CT: Yale University Press.

Index

About the Author

David Arditi is assistant professor of interdisciplinary studies at the University of Texas at Arlington. He received a PhD in cultural studies from George Mason University in 2012. While gigging as a drummer in Virginia, he became interested in the way that the labor of musicians is often overlooked in discussions of the recording industry. His research explores the relationship between music and technology and the way that relationship affects music, culture, and society. Arditi has published in *Popular Music & Society*, *Civilisations*, and the *Journal of Popular Music Studies*.